How Does Germline Regenerate?

CONVENING SCIENCE

DISCOVERY AT THE MARINE BIOLOGICAL LABORATORY

A Series Edited by Jane Maienschein

For well over a century, the Marine Biological Laboratory has been a nexus of scientific discovery, a site where scientists and students from around the world have convened to innovate, guide, and shape our understanding of biology and its evolutionary and ecological dynamics. As work at the MBL continuously radiates over vast temporal and spatial scales, the very practice of science has also been shaped by the MBL community, which continues to have a transformative impact the world over. This series highlights the ongoing role MBL plays in the creation and dissemination of science, in its broader historical context as well as current practice and future potential. Books in the series will be broadly conceived and defined, but each will be anchored to MBL, originating in workshops and conferences, inspired by MBL collections and archives, or influenced by conversations and creativity that MBL fosters in every scientist or student who convenes at the Woods Hole campus.

How Does Germline Regenerate?

Kate MacCord

The University of Chicago Press

Chicago and London

An open access digital edition of this book is available thanks to the
James S. McDonnell Foundation and the Marine Biological Laboratory.

The terms of the license for the open access digital edition are
Creative Commons Attribution-Non-Commercial-No-Derivatives 4.0
International License (CC BY-NC-ND 4.0). To view a copy of this
license, visit https://creativecommons.org/licenses/by-nc-nd/4.0/.

The University of Chicago Press, Chicago 60637
The University of Chicago Press, Ltd., London
© 2024 by Kate MacCord
Subject to the exception mentioned above, no part of this book may
be used or reproduced in any manner whatsoever without written
permission, except in the case of brief quotations in critical articles and
reviews. For more information, contact the University of Chicago Press,
1427 E. 60th St., Chicago, IL 60637.
Published 2024
Printed and bound by CPI Group (UK) Ltd, Croydon, CR0 4YY

33 32 31 30 29 28 27 26 25 24   1 2 3 4 5

ISBN-13: 978-0-226-83049-0 (cloth)
ISBN-13: 978-0-226-83051-3 (paper)
ISBN-13: 978-0-226-83050-6 (e-book)
DOI: https://doi.org/10.7208/chicago/9780226830506.001.0001

Library of Congress Cataloging-in-Publication Data

Names: MacCord, Kate, author.
Title: How does germline regenerate? / Kate MacCord.
Other titles: Convening science.
Description: Chicago : The University of Chicago Press, 2024. | Series:
   Convening science: discovery at the Marine Biological Laboratory |
   Includes bibliographical references and index.
Identifiers: LCCN 2023014976 | ISBN 9780226830490 (cloth) |
   ISBN 9780226830513 (paperback) | ISBN 9780226830506 (e-book)
Subjects: LCSH: Germ cells. | Somatic cells. | Cells.
Classification: LCC QL964.M33 2024 | DDC 571.8/45—dc23/
   eng/20230406
LC record available at https://lccn.loc.gov/2023014976

♾ This paper meets the requirements of ANSI/NISO Z39.48-1992
(Permanence of Paper).

Contents

INTRODUCTION | 1

1 Uncovering Assumptions That Have
  Shaped Germ Cell Science | 9
2 Backgrounding Conflicts within
  Germ Cell Science | 42
3 Challenging Assumptions in Germline Science | 61
4 Implications of Reenvisioning Germline
  Regeneration | 108

EPILOGUE | 123
ACKNOWLEDGMENTS | 125
NOTES | 129
BIBLIOGRAPHY | 139
INDEX | 161

Introduction

Infertility. Evolution. Heredity. Regeneration. What unites these disparate but important topics? Sex cells—like eggs and sperm—do. Scientists call these germ cells. They allow us to make more generations of ourselves. Because germ cells are the means by which we and all other sexually reproducing organisms procreate, they are tied to a wide variety of biological problems and questions. Consider, for instance, how species change over time—accreting changes over generations by passing heritable traits down through offspring. Evolution and heredity hinge on germ cells. Now consider that one in eight couples worldwide suffer from infertility. Some of these infertility issues involve the ability to make (or make properly) germ cells.

Now consider regeneration. Regeneration occurs in all living systems wherein a disturbance or injury to the system causes a response to repair what was lost or damaged. When we think of regeneration, we tend to think of it happening in organisms. A salamander loses its limb, and it regrows its limb. You cut your finger, and the skin regrows and heals. But regeneration can happen within any living system, whether it is an organism, a microbial community, or even an ecosystem. It can also happen in germ cells.

In a previous book in this series, historian and philosopher of biology Jane Maienschein and I laid out a brief history of thoughts about regeneration and raised questions about how we can understand regeneration across living systems by thinking about it as a systems-based phenomenon.[1] As we discussed, regeneration is a major focus for medical treatments and interventions and also has implications for things that range from our bodies' reactions to antibiotics to restoring the health of our fractured global ecosystems. Understanding regeneration, then—what it is, how it works, and how lessons learned from one system can be applied to another—is important.

In this book, I am concerned with germline regeneration. Let's take a moment to unpack what this means. *Germline* refers to a lineage of germ cells. Germ cells are sex cells: eggs and sperm. Scientists call these *oocytes* and *spermatozoa*, or (collectively) gametes. Beyond gametes, other types of germ cells occur in the body throughout the course of development, including but not limited to primordial germ cells and germline stem cells. Scientists consider these germ cells to constitute what they call a cell lineage. Being a cell lineage means that there is supposed to be a developmental history that connects all the germ cells from the early embryo to the gametes. Think of it as a family tree. All members of the tree are related through lines of descent, and the tree figure shows you what those relationships are. The same principle can be applied to germ cells, such that primordial germ cells that form early in

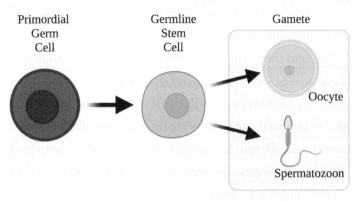

Germline (Cell Lineage)

FIGURE I.1 | Germline cell lineage. Germline is initiated when the primordial germ cells form. These eventually give rise to germline stem cells, which, through a series of intervening cell types, give rise to the gametes (oocytes and spermatozoa). Created with BioRender .com.

the embryo eventually give rise to germline stem cells, which sit within the gonads (reproductive organs) and give rise to gametes. Thus, all germ cells within an individual are considered members of a single lineage or family tree, and this cell lineage is commonly called germline (fig. I.1).

Many scientists tend to think of germline as something special that allows all sexually reproducing organisms to create new generations. They also tend to think of this cell lineage as distinct—as a separate lineage of cells that doesn't intermingle with the rest of the cells in the body—called *somatic* cells. Somatic cells include everything from the red blood cells coursing through your veins to the neurons allowing you to read this book. Scientists also consider somatic cells to have lineages, but these cells are thought to be fundamentally different from germ cells because only germ cells are the bearers of our heredity. Germ cells bridge the divide between us and the next generation, carrying information from our bodies into the future, whereas somatic cells do not.

I've said that I am concerned throughout this book with germline regeneration. In theory, germline regeneration means that if any of the types of cells that make up germline are damaged or removed, they will be repaired, or more likely, replaced, resulting in fully functioning gametes that can give rise to progeny. In practice, germline regeneration is quite a bit more complicated and controversial.

Why is germline regeneration complicated and controversial? *Because many scientists say it is not supposed to happen except in very limited circumstances.* What does this mean? After all, human males produce millions of sperm daily. Human females, on the other hand, are believed to have a limited number of oocytes that have already formed at the time of their birth (although there is growing reason to doubt this).[2] When I say that many scientists think that germline only regenerates under very limited circumstances, I mean that scientists largely (but not entirely) hold a particular view of germline regeneration: germ cells can only regenerate from other germ cells. Let's think about what this means by looking at a hypothetical example. Let's say that every germ cell is removed from an adult male mouse—all of the germline stem cells and all of the germ cells in varying stages of differentiating into spermatozoa are removed. In this scenario, germline is not supposed to regenerate within this mouse because there are no germ cells available to regenerate what was lost. The result is an infertile mouse. Germline regeneration, then, is supposed to be impossible in the absence of germ cells.

In order to envision how limited this view of germline regeneration is, let's compare it to another form of regeneration—that of the salamander limb. When a salamander loses its limb, its body forms a layer of cells over the wound. This cell layer is called a *blastema*. The blastema is composed, in part, of cells from the area surrounding the injury that have dedifferentiated and become stem cells. *Dedifferentiated* means that the cells have changed from one type of cell (like a muscle cell) to a less specialized type of cell (like a stem cell). These stem cells will help produce the tissues and structures of the regenerating limb. In salamander limb regeneration, then, a variety of specialized cells from different cell lineages will ultimately contribute to the new limb, whereas in germline regeneration, only one cell lineage (germline) is supposed to contribute to the new germ cells.

Thinking about germline regeneration as restricted to regeneration only from cells within its own lineage relies on scientists being able to do three intertwined things that build on one another. First, they have to claim that they can accurately and reliably distinguish germ cells from all of the other cells in the body (somatic cells). Second, they need to be able to say definitively that the germ cells maintain a continuous cell lineage such that each germ cell in the body can only be traced through development to other germ cells. Third, they need to be able to state that under no circumstances can somatic cells become germ cells. Each of these is an empirical claim that underlies the limited view of germline regeneration outlined above, and scientists need to be able to affirm each of these claims in order to believe that germline can only regenerate from germ cells. And yet, these are empirical claims that are not often explicitly questioned within germline research—it is most often assumed that the answer would be yes. We can therefore think of these claims as assumptions—as things that many scientists take for granted within their research programs. We should occasionally prod things that we take for granted, in life as well as in science, and see whether they are true. After all, the things we take for granted shape how we view the world, condition us to perceive things in certain ways, and can blind us to alternative possibilities.

This book is the outcome of a collaboration between me (an historian and philosopher of science) and B. Duygu Özpolat (a germline biologist). Understanding the ideas and practices behind Özpolat's work on germline regeneration led me to pinpoint the assumptions outlined above and question their origins and validity. Recent science thus motivates my driving question for this book: *How does germline regenerate?*

Answering this question in a meaningful way requires me to take an unorthodox approach. First, I explore each of the three assumptions that underlie germline regeneration, focusing on how they arose within history, whether they were valid at their origins, and how they became embedded within our understanding of germ cells, germline, and germline regeneration. I begin with the third assumption—that under no circumstances can somatic cells become germ cells—because it is historically the oldest. This historical and philosophical approach gives insights into why and how these three key assumptions are held widely by scientists today. Next, I challenge the validity of each of the assumptions by dissecting their current meanings and what evidence supports them as empirical claims within recent science. In the end, I bring these threads of history, philosophy, and science together to show how assumptions about the nature of germ cells and germline have massive repercussions for issues, such as human genome editing, that sit at the intersection of science and society.

This book, then, is about germline regeneration—how many scientists assume it works and how we can reenvision its working. It is also about how science works, how history shapes current science, how science can shape the practice of history, and how things that we take for granted within science can have far-reaching effects.

1   Uncovering Assumptions That Have Shaped Germ Cell Science

Our understanding of biological problems has been shaped by the history of thinking about these problems. The history of science is not a series of eureka moments; it is a series of conflicts, competing concepts, and conceptual shifts that accrete over time. Underlying all of this are assumptions. Assumptions are an inherent part of science; they are the theories, methods, and empirical claims that are deployed unchallenged within research programs. Scientists must take things for granted when they observe, experiment on, or make policies about the natural world. For example, when scientists observe the migration of primordial germ cells in an embryo, they have to assume that the methods that they use to identify those primordial germ cells are accurate. They can assume this because a long history of identifying primordial germ cells makes it seem like a safe assumption to make.

Assumptions are not, by their nature, good or bad. They are simply a part of the routine practice of science. Nonetheless, we should still examine them and ensure that the evidence and reasoning behind them is sound, because they shape the ways in which we understand the world. One assumption that we need to revisit is the one inherent in the concept called the Weismann Barrier. The Weismann Barrier is supposed to act as an impediment between germ cells and somatic cells, and it establishes a definite relationship between the two: germ cells can give rise to somatic cells, but somatic cells can never become germ cells. One assumption leads to another: the idea that germline can only regenerate from germ cells. Assumptions can therefore canalize our views and condition us to think about the world in particular and exclusionary ways.

One way of prodding assumptions in science is to look to history—to see how these assumptions became embedded in the first place and what evidence there was to support them. By tracking the origins of assumptions and how they have changed over time, we can begin to understand how science works and how the concepts and tools that current scientists rely on became established. Using this historical perspective, we can also see problems with assumptions that current scientists hold, begin to explain how these problematic assumptions took root and became disconnected from their historical, evidential basis without reassessment, and point to areas where further scientific scrutiny is necessary. In other words, looking at history can shape current science.

Throughout this chapter, I examine the historical thinking behind the nature and origins of germ cells and look at how scientists established the Weismann Barrier in historical context. The different lines of this narrative are thus framed by my interest in particular scientific developments that led to the idea of the Weismann Barrier and exclude the full breadth and richness of historical detail surrounding the individuals, concepts, and research programs at play. This history is one of shifting assumptions about the nature of germ cells, the fracturing of the problems of heredity and development into separate fields, and the conflation and reinterpretation of a theory meant to explain both problems.

DISCOVERY AND ORIGIN OF GERM CELLS

In the seventeenth century, scientific thinkers did not have a cohesive concept of cells, let alone germ cells. At this time, newfangled technology called microscopes first allowed researchers to see cells, to question how different types of cells were related, and to debate their importance for plants and animals. The origin of thinking about germ cells is the origin of thinking about gametes. The first person to document what modern researchers would call a gamete was the famed Dutch microscopist, Antonie van Leeuwenhoek, one of the earliest people to document observations of cells. In 1677 Leeuwenhoek observed ejaculate using microscope lenses that he had crafted.[1] He saw small things with ovoid heads and thin, undulating tails, which he called "animalcules" and later "spermatozoa" (fig. 1.1). Leeuwenhoek's work established that spermatozoa were cells. What exactly it meant for sperm to be cells was unclear in Leeuwenhoek's time; it was more than a century and a half before cell theory established that cells were individual units of living organisms.

Leeuwenhoek's discovery helped to prompt questions

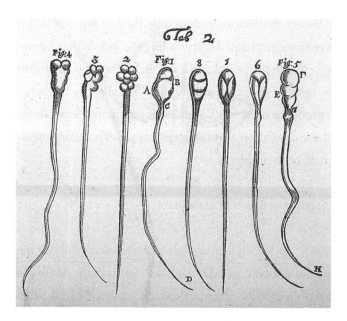

FIGURE 1.1 | Antonie van Leeuwenhoek's sketches of spermatozoa. Table 2 from "Observationes D. Anthonii Lewenhoeck, De Natis E Semine Genitali Animalculis," *Philosophical Transactions* 12 (1678): 1040–46.

about whether mammalian females also had gametes, or ovi or eggs, as they would have been called at the time, and if so, what they looked like and where they could be found.[2] Mammalian ova proved far more elusive than spermatozoa, because they are hidden away within the ovaries and uterus. The first observation of ova was made 150 years after Leeuwenhoek's identification of spermatozoa. In the spring of 1827, while dissecting the ovaries of a dog, Estonian scientist Karl Ernst von Baer observed ova.[3] Through a series of dissections, Baer confirmed the presence of ova in other mammals and was able to declare that, like males, females created gametes.

By 1827 then, scientists had established that spermatozoa

and ova were individual cells. With the advent of cell theory over the next three decades, and its tenet that all cells must come from preexisting cells, the question became, Where did these spermatozoa and ova come from? Scientists recognized that gametes formed inside of the testes and ovaries, but they wanted to understand how this worked and whether progenitors of these gametes formed before testes and ovaries developed. In 1870, two major works appeared that separately surveyed the origins of gametes in vertebrates and invertebrates.[4] The anatomist Heinrich Wilhelm Gottfried Waldeyer, working in Poland,[5] conducted the work on vertebrates. Waldeyer is best known today for coining the terms *neuron* and *chromosome*, but in the late nineteenth century, Waldeyer was a famous and eclectic anatomist who looked at everything from the development of teeth to the anatomy of neurons and even the structures inside of cells. In 1870, Waldeyer became interested in understanding where, during development, the cells that give rise to the gametes originate. He looked at the development of ova across several groups of vertebrates and concluded that ova and ovaries and spermatozoa and testes arose from two different types of epithelium (a kind of tissue) within the developing embryo. Because vertebrate embryos possessed both of these types of epithelium despite their final sex, Waldeyer concluded that embryos are initially hermaphroditic.

Meanwhile, Edouard Van Beneden, an embryologist and cytologist working in Belgium, conducted the work on invertebrates. He argued that the cells that gave rise to gametes of both sexes originated at the earliest stage of development, not once the embryo had progressed to the stage where it contained different tissues, as Waldeyer concluded. And, because the cells that gave rise to gametes originated so early in van Beneden's estimation, they could not be traced to a specific tissue, as Waldeyer had done. Thus, these works came to conflicting conclusions about when and where the cells that gave rise to gametes originate during development. Despite the fact that neither Waldeyer's nor Van Beneden's work was very comprehensive by modern standards, they helped to spark an interest in the origins of germ cells.[6]

THE NINETEENTH CENTURY: EVOLUTION AND HEREDITY

In the late nineteenth century, thinking about evolution imbued research within biology: researchers questioned how evolution worked and the kinds of evidence that supported the theory, and they used it as a backdrop for building new ideas about the natural world. All of this followed English naturalist Charles Darwin's publication in 1859 of *On the Origin of Species*, in which he laid out his theory of evolution by natural selection.[7] Although Darwin's theory was not universally accepted within the scientific community in the decades following its publication, many scientists embraced Darwin's ideas or at least the central idea that life evolves in some way.[8]

The theory Darwin proposed relied on his belief that organisms can pass traits between generations, that those traits can vary within generations, and that natural processes "select" which traits make organisms better suited to their environments. Because of this, scientists began increasingly to focus on the problem of heredity, seeking to understand how traits get transmitted through generations and how they could do so in a way that allowed for the variation that was paramount to natural selection's operation.

One of the prevailing theories of trait variation in the nineteenth century was the idea of inheritance of acquired characteristics. This notion was proposed by French naturalist Jean-Baptiste Lamarck in 1809, who held that organisms could acquire small variations through the use or disuse of parts throughout their lifetimes.[9] The standard example given for this idea is the giraffe neck: each generation of giraffe ancestors extended their necks just a little bit farther to reach for food, and those small changes accumulated over time to produce the long-necked organism we now know. While Lamarck's theory of inheritance of acquired characteristics wasn't a theory of heredity *per se*, it provided a framework for understanding how traits could vary within populations over time.

Darwin himself was quite partial to the idea that the use or disuse of parts could reshape an organism and get passed through to the next generation. In 1868, he proposed the theory of pangenesis, his own theory of heredity, in which he hypothesized that each part of the body continually emits its own type of small, organic particle.[10] Darwin called these particles *gemmules*. Gemmules, Darwin thought, circulate freely within the body, and given proper nutrition, they can turn into the tissues within the body. His theory was vague about which tissues gemmules could form—that is, whether they formed only the tissues from which they were derived or could form other tissues as well. The gemmules, for Darwin, contained information about the parts of the body from which they were derived: thus, they carried information for the organism's traits.

Darwin believed that gemmules derived from all over the body coalesced within the gonads and from there contributed to spermatozoa and ova, and thus reproduction. Gemmules, then, were the material of heredity, and when they formed an embryo, the information contained within the gemmules would direct the development of the organism. When gemmules from two parents coalesced to form a single organism, their material and the information they carried were thought to blend together. This blending of gemmules explained why offspring are not exact replicas of a single parent but instead tend to have features that blend the appearance of both parents.

Darwin's theory of pangenesis also accounted for the origins of variations that form the basis of evolution. It did so by appealing to a version of the inheritance of acquired characteristics that Lamarck had made famous. Each gemmule, Darwin believed, could acquire new characteristics by responding to changes within its environment. Gemmules that were derived from the skin, in Darwin's account, if constantly exposed to light, could pass on a darker skin pigment to offspring than the parents had. The same, Darwin believed, was true for things like muscularity—any and every trait was subject to the effects of the environment and could be modified during life and then passed on in a modified form to offspring.

Darwin's theory of evolution explained how traits could be shaped over time within populations, but it gave no mechanism for how (1) variation arose in the first place, or (2) how traits moved between generations. In crafting the theory of pangenesis, Darwin was conscious of these holes in his theory of evolution, and actively tried to fill them.

Darwin's theory of pangenesis met with mixed reviews.[11] Some scientists and the popular press lauded his theory,[12] not necessarily because it gave evidence for how heredity worked, but because it filled in the gaps in his theory of evolution. Pangenesis provided a mechanism by which variation originated and got passed onto offspring. Some scientists, including Dutch botanist Hugo DeVries and American zoologist William Keith Brooks, proposed variations on the theory of pangenesis and sought experimental evidence to support it.[13]

Meanwhile, many scientists were hesitant to embrace Darwin's theory of heredity, largely because there was no direct evidence to support it. Nobody could see the gemmules directly, for example. Some were also reluctant to embrace a theory of heredity in which the origin of traits, and thus evolution, relied so heavily on the inheritance of acquired characteristics—a notion that was also not observable and that was heavily debated within the scientific community. One scientist who held such a view of Darwin's theory was the German embryologist and evolutionary theorist August Weismann, an eminent scientist in the late nineteenth century who, as historian Frederick B. Churchill suggested in his comprehensive biography of Weismann, sought to bring together some of the biggest phenomena of life, such as heredity, development, cells, and evolution, under a big theoretical umbrella.[14]

AUGUST WEISMANN

August Weismann studied medicine in Germany, and upon receiving his degree, he traveled around Europe, working as a personal and military physician. While working as a physician, Weismann continued his studies, conducting research on natural history and addressing questions of organismal development. As Weismann delved into biological research, he became convinced that Darwin's theory of evolution by natural selection was the only theory that fit with all of the evidence from nature.

Although Weismann believed Darwin's theory of evolution by natural selection was correct, he also quickly became convinced that the inheritance of acquired characteristics, which Darwin and others had supported within their theories of heredity, was untenable. In 1888, Weismann delivered a lecture at the Association of German Naturalists in Cologne, Germany, in which he described the results of an experiment intended to test the supposed heritability of acquired characteristics.[15] In this experiment, Weismann cut off the tails of 901 mice and their offspring across five generations. Weismann reasoned that if acquired characteristics were heritable, these mice should eventually produce offspring with no tails. This did not happen, leading Weismann to conclude that acquired characteristics could not be inherited.

Weismann's rejection of the inheritance of acquired characteristics also arose from his own research on development and heredity. By the time that he delivered the results of his experiments on mouse tails, he had already begun to formulate a theory about how heredity works that rendered the inheritance of acquired characteristics impossible.

In the late 1870s, Weismann began to study how germ cells originate and develop within several species of small, predatory marine invertebrates closely related to jellyfish, called hydrozoans. He was working within the framework of Darwinian evolutionary thinking and really wanted to understand how heredity works. For Weismann, understanding heredity meant not only figuring out how traits can move between generations, but also how, during development, cells could differentiate into all parts of the body, becoming the traits that we recognize within individuals. In other words, Weismann wanted to ground his understanding of heredity within organismal development, and whatever theory he generated from this work had to account for not only the movement of traits between generations, but also how those traits were generated within an organism during its development. In 1883, Weismann published his findings on the origins of sex cells in hydrozoans, which became the basis for his theory of heredity.[16] Over the next nine years, Weismann refined his theory, and in 1892 he published a full account of his understanding of heredity in *The Germ Plasm*.[17]

Weismann's germ plasm theory of heredity is rather complicated; for my purposes it's important to understand three things. First, he believed that heredity had a material basis that was contained within the nuclei of all cells, which he called "germ plasm." Second, germ plasm was a theoretical substance, meaning that he did not have direct evidence that it existed, although his argument for its existence was well supported. For

Weismann, the chromosomes were the structures that probably contained the germ plasm. Weismann could not confirm that chromosomes contained the germ plasm, but evidence about the ways in which chromosomes behave during cell division led him to believe that they probably did. Chromosomes had only recently been recognized as having individuality and continuity across cell divisions when Weismann published his full theory.[18] Third, the germ plasm contained component parts that were arranged in a particular, set architecture that was significant for the cells and for Weismann. Let's briefly take a look at this germ plasm architecture.

Weismann conceived of the germ plasm as containing four levels of components, arranged in a hierarchical way: *biophors, determinants, ids,* and *idants* (fig 1.2). At the lowest level of the hierarchy were biophors, tiny units that made up all of the structures of cells. Biophors were collected into slightly larger units called determinants. Determinants were supposed to give cells their specific identities because, according to Weismann, only one determinant could be active within a cell that had been fully differentiated. Determinants, then, defined cell types and guided their various functions. Aggregates of determinants were called ids. Each id was supposed to contain enough hereditary material (i.e., biophors and determinants) to produce a whole new organism. Together, groups of ids made up each idant, which likely corresponded to the chromosomes.

So, how did Weismann envision this germ plasm, with its hierarchical structure, working as the material basis of heredity? We need to recall first that Weismann grounded his think-

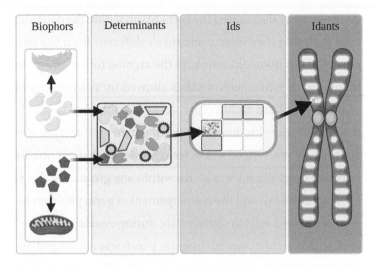

FIGURE 1.2 | Architecture of Weismann's germ plasm. Biophors are the lowest level of the germ-plasm hierarchy, and these make up all of the cellular structures. The biophors are collected into larger units called determinants, which give cells their individual identities. Determinants are aggregated into ids, which contain enough hereditary material to produce a new organism. Ids are grouped together into idants, which probably correspond to chromosomes. This image is based on figure 1 in Bline, Goff, and Allard, "What is lost?" Created with BioRender.com.

ing in organismal development: his vision of germ plasm accounted both for how traits arise during development *and* for how traits pass between generations. As an organism develops following fertilization, cells divide, and through successive rounds of cell divisions, cells differentiate. When cells divide, their nuclei divide, and Weismann envisioned that nuclear division (and thus division of the germ plasm) could happen in two different ways. The first way he called *homeokinesis*. During homeokinesis, the nucleus of the cell would divide in a way that replicated the architecture of the germ plasm exactly within the nucleus of each daughter cell. The second way he called *heterokinesis*. During heterokinesis, the nucleus would

CHAPTER ONE

divide in a way that altered the germ plasm within the daughter cells such that they were *qualitatively* different from the parent cell. Thus homeokinesis kept the architecture of the germ plasm intact, while heterokinesis allowed for rearrangement of the germ plasm architecture. As cells continued to divide throughout development, the germ plasm architecture was continually rearranged via heterokinesis, so that eventually only one determinant was active within any given cell. This is how heterokinesis and the rearrangement of germ plasm architecture allowed cells to differentiate during development.

After the architecture of the germ plasm was rearranged following heterokinesis, Weismann called it "idioplasm," and all cells that contained idioplasm were called "somatic cells." He used this name change in order to distinguish the altered form of the germ plasm that led to new cellular identities (idioplasm) from the unaltered form of the germ plasm that would contribute to reproduction. Weismann called cells that contained this unaltered germ plasm "germ cells." He acknowledged that germ cells were not always identifiable at the beginning of development, so he devised the idea of the "germ-track" (the German word Weismann used was *keimbahn*). A germ-track is the lineage of cells that leads from the single cell of the fertilized egg to the emergence of the first germ cell (what Weismann called the *urKeimzelle*, from the German words *ur* for "original" or "primary," *keim* for "germ," and *zelle* for cell). The length of the germ-track varied between species. At the time, Weismann was unable to see exactly which cells in the different species he observed contributed to the germ-track.

This is where Weismann's theory gets even trickier—he

considered the cells within the germ-track to be somatic cells. How did this work? The somatic cells that were committed to the germ-tracks carried within them a full complement of unalterable germ plasm that was unaffected by the differentiation and physiological changes of the somatic cell. Weismann explained the ability of organisms to regenerate with a similar logic; parts of organisms that were the most subject to injury contained accessory, unalterable germ plasm. This accessory germ plasm was established within the cells that would eventually give rise to these regeneration-enabled parts at the earliest stages of development. The presence of this accessory germ plasm allowed cells that contained it to regrow missing parts, like salamander limbs, because the germ plasm could give those cells the ability to produce any cell type necessary to replace what was lost. Germ cells, to Weismann, were the cells at the end of the germ-track that (1) only contained unaltered germ plasm, (2) gave rise to daughter cells that only contained unaltered germ plasm (via homeokinesis), and (3) were the only cells capable of contributing to reproduction.

Early in his formulation of the germ plasm theory of heredity, Weismann thought that the germline was specified during the very first cell divisions of development: the first cleavage of the fertilized egg resulted in one germ cell and one somatic cell. While this somatic cell would eventually form all the other cells in the body, the germ cell would exist in a continuous lineage throughout development, giving rise to all other germs cells which were then capable of participating in reproduction. Others shared these beliefs. Beginning in the mid-1870s, at least two other German scientists, Gustav Jaeger and Moritz

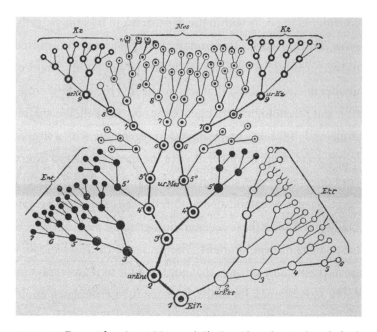

FIGURE 1.3 | Figure 16 from August Weismann's *The Germ Plasm*, showing the early development and germ-track in a species of roundworm (*Rhabditis nigrovenosa*) up to the twelfth generation of cells. The numbers indicate the various generations of cells. The cells of the germ-track are connected by thick black lines. The germ cells are black with white nuclei. Mes = mesenchyme, Ekt = ectoderm, Ent = endoderm, urKz = primordial germ cells, Kz = germ cells. (Weismann, *Germ Plasm*, 196)

Nussbaum, came to the conclusion that there was a continuity of germ cells from the fertilization of the egg throughout development and into the production of the ova or spermatozoa that would lead to the next generation.[19] By the time Weismann published his full account of germ plasm theory in 1892, however, he had recognized that, in fact, the germ cells could be established during later stages of development. We can see this in figure 1.3, which shows a diagram that Weismann published in *The Germ Plasm*. The germ cells (indicated by "Kz," from the German for germ cell, "*Keimzelle*") in this diagram aren't separated from the other cells until the ninth round of cell division. The presumption here is that the germ plasm architecture was maintained unaltered within a group of somatic cells throughout those earlier divisions. By the time that Weismann published his full account of germ plasm theory in 1892, then, his thinking had shifted away from *germ cells as continuous* to *germ plasm as a continuous material* that existed from the fertilized ovum through the germ-track to the first germ cells and then throughout the ova or spermatozoa, which then gave rise to the next generation. The germ plasm was thus continuous throughout development and linked generations of organisms. The continuity of the germ plasm was paramount to Weismann; it had to be continuous for the germ plasm theory to account for development and heredity.

Weismann's ideas about the germ-track and the germ plasm were a response to and rejection of the concept of inheritance of acquired characteristics, and especially Darwin's theory of pangenesis. By recognizing the complete, unaltered and continuous germ plasm as the sole basis of heredity, Weismann ensured that no mutations or characteristics acquired by idioplasm (altered germ plasm) could contribute to reproduction because idioplasm could not, by definition, be part of germ cells. Thus, only mutations within germ plasm could get passed on to future generations.

Weismann, through his germ plasm theory, thus proposed that there was a continuity of a material, called germ plasm, that existed within and between generations of organisms. The qualitative nuclear division of this substance (thereafter, idioplasm) accounted for the distinction between germ cells and somatic cells and allowed somatic cells to differentiate into all of the different cellular identities that make up the body. Weismann's germ plasm theory was very controversial when it was published and came under intense criticism on two fronts.

First, all but a handful of Weismann's contemporaries resoundingly objected to the notion of qualitative nuclear division of cells (heterokinesis). Experiments on developing embryos in the years following Weismann's publication of the germ plasm theory definitively showed that there was no qualitative nuclear division during development.[20] Second, Weismann's contemporaries objected to the theoretical and abstract nature of the germ plasm. There was not, after all, direct, experimental evidence to support the existence of biophors, determinants, idants, and ids, but there was a great deal of evidence to support the logic of his germ plasm theory. For instance, by the time Weismann published the germ theory, scientists knew that there was a material basis for heredity. They knew that chromosomes came apart during cell division and that cells differentiated during development. In many respects, Weismann's theory fit the phenomena of development and heredity, and his critics largely acknowledged this. So, while some of Weismann's strongest critics objected to his theory on the grounds that there wasn't direct evidence to support it, what many really objected to was the abstract nature of the theory—that it could not yet, with available technology, be experimentally determined whether Weismann's germ plasm components existed or, if so, whether they operated in the ways that he claimed. Therefore, they thought it was largely untestable and thus unscientific. We can see this thinking in a scathing review of the germ plasm theory written by American cytologist Edmund Beecher Wilson: "From a logical point of view the Roux-Weismann theory is unassailable. Its fundamental weakness is its quasi-metaphysical character, which indeed almost places it outside the sphere of legitimate scientific hypothesis . . . Such an hypothesis cannot be actually overturned by an appeal to fact. When, however, we make such an appeal, the improbability of the hypothesis becomes so great that it loses all semblance of reality."[21]

In this quote, Wilson, like many of his contemporaries, lumps Weismann's germ plasm theory together with the ideas about development and differentiation developed by the German embryologist, Wilhelm Roux. Roux also believed that cell division during development resulted in a qualitative difference amongst cells, but other aspects of his thinking differed from Weismann, as historian Fred Churchill has explained in depth.[22] Thus, Weismann's theory was received skeptically by his peers and largely discredited. And yet, rejection of his germ plasm theory does not mean that all of Weismann's ideas got swept into the ash heap of history.

ESTABLISHING THE WEISMANN BARRIER

Just a few years after Weismann published his full account of the germ plasm theory, Edmund Beecher Wilson, one of the scientists who had so critically objected to Weismann's theory made parts of Weismann's thinking integral to his textbook, *The Cell in Development and Inheritance*.[23] Wilson studied biology in the United States and is a preeminent figure in the history of cytology (the field that we now call *cell biology*). He is generally considered to be one of the founding figures of modern American experimental biology.[24] His text, which highlighted his prowess in experimental cell research, proved popular and highly influential, going through three editions between 1896 and 1925. It was considered essential reading for Anglophone scientists studying cells in the early twentieth century, and the ideas that Wilson proposed thus became widespread throughout the scientific community.[25]

Throughout all of these editions, Wilson grounded much of his thinking in some aspects of Weismann's ideas about *germ cells*. In particular, he thought there was a distinction between germ cells and somatic cells, that there are cell lineages that give rise to germ cells through development, and that germ cells enable heredity. But Wilson made substantial modifications to Weismann's understanding of germ cells and heredity, and completely rejected the idea of Weismann's germ plasm and his notion of qualitative reduction to delineate germ plasm from idioplasm, and thus, germ cells from somatic cells. Wilson was far more focused on explaining development and how cells divide and differentiate over time, and understanding heredity was a by-product of this. Meanwhile, Weismann was focused on understanding heredity, and explaining development and differentiation was a by-product to him. Moreover, while Weismann sought a unifying theory of heredity that could account for things like differentiation, development, and even regeneration, Wilson hewed much more closely to the scientific evidence and did not propose a generalized theory of development, let alone heredity.

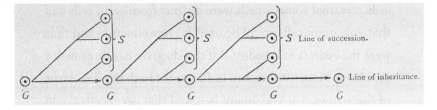

FIGURE 1.4 | Figure 4 from Edmund B. Wilson, *The Cell in Development and Inheritance* (p. 11). The figure is labeled "Diagram illustrating Weismann's theory of inheritance." The figure legend in the original text reads, "G. The germ-cell, which by division gives rise to the body of soma (S) and to new germ-cells (G) which separate from the soma and repeat the process in each successive generation."

While Wilson rejected the germ plasm theory, he respected Weismann's success in uniting evolutionary theory with cell theory, and it was at this juncture that Wilson situated his thinking. As he explains in his text, "For aside from the truth or error of his special theories, it has been Weismann's great service to place the keystone between the work of the evolutionists and that of the cytologists, and thus to bring the cell-theory and the evolution-theory into organic connection. It has been my endeavor to treat the cell primarily as the organ of inheritance and development; but, obviously, this aspect of the cell can only be apprehended through a study of the general phenomena of cell-life."[26] Out of his respect for Weismann's ability to connect evolutionary theory directly to cell theory, Wilson introduced a diagram to show his readers how Weismann's theory of heredity worked (fig. 1.4).

Wilson's diagram is a simplified abstraction of Weismann's thinking about germ cells and somatic cells, but it indicates the relationship that Weismann thought existed between them to a certain extent.[27] Weismann did think, as Wilson's diagram indicates, that somatic cells were distinct from germ cells and that only germ cells could be used for reproduction and thus were the bearers of heredity. But this diagram has three major problems. First, the diagram indicates through the directions of the arrows that Weismann believed that germ cells could not be derived from somatic cells. We have already seen that Weismann did not believe this; rather, he thought that somatic cells were a part of the germ-track and thus gave rise to germ cells. Second, the diagram and description appear to indicate a continuity of germ *cells* within and between generations. As detailed above, Weismann believed in a continuity of germ *plasm*, not of germ cells. The germ cells could arise well after the first cleavage of the embryo, and so could not be continuous between generations. The germ plasm, meanwhile, was the driver of heredity and development, and therefore had to be continuous. Indicating a continuity of germ cells gave the strong impression that the germ cells were a separate cell lineage from the somatic cell lineages. Third, the diagram is missing the most crucial aspect of Weismann's thinking—the germ plasm. The germ plasm was the distinguishing feature for Weismann, not the cells themselves—those were the by-products of changes in the germ plasm (or subsequent idioplasm).

Wilson's diagram is interesting in how it both represents and misrepresents Weismann's ideas about relationships between germ cells and somatic cells, his theory of inheritance, and his concept of continuity. This reformulation of Weismann's theory of heredity is important for my history because it was featured in a very popular and important textbook. The popularity of Wilson's book and the esteem with which he was held meant that this reformulation of Weismann's theory into a simple relationship between germ cells and somatic cells became widespread among students and professionals. Our post-Wilson description of the relationship between germ cells and somatic cells became known as the "Weismann Barrier." This so-called barrier is supposed to prohibit somatic cells from becoming germ cells, leaving us with a simplified understanding of the relationship between germ cells and somatic cells: germ cells can give rise to somatic cells, but somatic cells can never become germ cells. This is a very non-Weismannian notion. And yet it burrowed deep into biological thinking, associating Weismann's name with a vision he never held.

THE WEISMANN BARRIER ENTRENCHED

By 1900 many questions remained about germ cells: Where and how did they originate during development? How might germ cells within an individual be related to each other? How did germ cells operate within heredity? Before we move on to how scientists thought about and sought to resolve these questions during the early twentieth century, we need to understand how scientists were thinking about cells and their relationships with one another throughout development during the late nineteenth century. In other words, we need to understand the thinking behind conceiving of cells as lineages in which the descent of cells from the first cleavage of the fertilized egg could be tracked through the processes and timeline of development.

As we saw with Weismann's thinking about the germ-track

in his 1892 text *The Germ Plasm*, scientists interested in development during the late nineteenth century had begun to think of cells in ways similar to how scientists interested in evolution thought of species—as units with a traceable lineage of descent. While scientists largely traced the descent of species during this time by inference from morphological characters and biogeography, researchers could observe directly the descent of cells within a developing embryo. In 1878, American zoologist Charles Otis Whitman, who later became the first director of the famed Marine Biological Laboratory in Woods Hole, Massachusetts, published the first cell lineage study.[28] In this work, Whitman traced the division and development of each cell from the single-celled fertilized eggs of a group of leeches called *Clepsine* all the way through the origins of the different germ layers. Germ layers are rudimentary tissues within the early embryo that give rise to all the other tissues and organs.

As historian Jane Maienschein has shown, a group of young American scientists—including Edmund Beecher Wilson—followed in Whitman's footsteps in the 1890s, taking up the work of elucidating cell lineages in other marine and aquatic organisms.[29] Many of these scientists conducted this work at Whitman's Marine Biological Laboratory. These cell lineage studies demanded painstaking work. They required scientists to collect thousands of organisms, to sit for hundreds of hours at microscopes lit by the sun or candlelight, to watch as cells divided, and to preserve and stain tens of thousands of embryos in various stages of development. Those who completed these cell lineage studies showed how a single, fertilized cell could be traced through thousands of cell divisions into the multicellular organs and tissues that make up an organism. Conversely, they showed how, with a lot of effort and patience, scientists could trace individual organs or even cell types back to their developmental origins. This kind of thinking applied to germ cells, and by the turn of the twentieth century, scientists were thinking about germ cells in terms of their cell lineages. The cell lineage of germ cells became known as the germline, and this concept embodies all the different types of germ cells that exist throughout an organism's life history—primordial germ cells, germline stem cells, gametes, and so on—although knowledge about some of these cell types and the terms for them did not come until much later.

By 1900, the methods and ideas of understanding development by tracing cell lineages were embedded within embryology (the field that we now call *developmental biology*), even as many of the scientists who did the initial cell lineage studies, like Whitman and Wilson, had moved on to other ideas and experiments, and cell lineage studies became less and less widespread. At the same time, study of the problems of heredity and the problems of development, which Weismann had tried to unify in his germ plasm theory, began to diverge. Around 1900, a small group of scientists interested in questions of heredity rediscovered Czech monk and naturalist Gregor Mendel's work on pea plants, which helped spur the birth of the field of genetics. At the same time, researchers in embryology shifted their focus from describing and comparing developing embryos among species to crafting experiments that allowed them to intervene in development. Cutting or shaking apart early embryos from frogs, fishes, sea urchins, and other organisms became common practice by 1900, and scientific work on embryos moved away from the creation of big, overarching theories like those Weismann had pursued, to experimentally driven and hypothesis-driven research.[30] As these scientists pursued their quest to understand development in terms of cells and their interactions, a sometimes-overlapping but increasingly separate group of scientists pursued questions about genes and heredity. Questions about heredity and development thus came to reside in separate and increasingly divergent fields, genetics and embryology, respectively. And, while the field of genetics absorbed questions about the hereditary nature of the germline, embryology absorbed questions about the origins and developmental continuity of germline.

This fracture of problems like heredity and development into distinct fields is important for our purposes because scientists within these fields developed very different conceptions of the Weismann Barrier, germ plasm continuity and germline, and understanding germ cells during the early twentieth century. As embryologists John Normal Berrill and Chien-Kang Liu put it in 1948, "To many geneticists it [germ plasm continuity] still seems to have an odor of sanctity, [but] to most embryologists, it has an old-fashioned association with what are now regarded as problems or phenomena of development pure and simple."[31] When Berrill and Liu use the term *germ plasm continuity*, they mean to conflate Weismann's concept of the germ plasm (which he thought was continuous) with Wilson's interpretation of Weismann's theory, which made the germline (cell lineage) continuous. Thus, Berrill and Liu may say "germ plasm continuity," but what they mean is *germline continuity* and thus the Weismann Barrier.

Geneticists in the early twentieth century were keen to adhere to the idea of the Weismann Barrier. It provided a convenient means of excluding the inheritance of acquired characteristics from concepts of heredity (and later, a means of connecting genetics, heredity, and evolution) and indicated that genetic material could remain stable, despite external conditions. Embryologists, meanwhile, were much more diverse in their reactions to the Weismann Barrier and the attached concept of germ plasm continuity at the time.

Gregor Mendel is often credited as being the father of genetics because of his 1866 paper on inheritance of traits in pea plants.[32] When his paper was published, it was largely ignored by the scientific community. That changed when Mendel's work was "rediscovered" around 1900 by a small group of scientists, including Dutch botanist Hugo de Vries, German botanist Carl Correns, Austrian botanist Erich von Tschermak-Seysenegg, and English biologist William Bateson.[33]

What is important to understand here is that around 1900 the problem of inheritance and questions about how it operates were beginning to coalesce within the incipient field of genetics. At the turn of the twentieth century, genetics was very different from how we understand it today. Scientists at this time had no concept of DNA, base pairs, or even genes. In fact, the term *gene* was coined by Danish botanist, Wilhelm Johannsen, in 1909 to describe the unit of hereditary information, and in that same year William Bateson coined the term *genetics* to describe the study of inheritance and variation. At this time, the field of genetics was diverse in its objectives and methods, ranging from the biometricians who relied on statistics to understand heredity and evolution to the Mendelians who experimentally investigated the appearance and variation of organismal traits and increasingly focused on the chromosomes.[34] In 1915, American biologist and future Nobel Prize winner Thomas Hunt Morgan and his collaborators published a book called *The Mechanism of Mendelian Heredity*, in which they laid out the principles of heredity in terms of genes (which were still theoretical structures at this point) and chromosomes.[35] This book, along with the research of others, helped to give a solid foundation to what became transmission genetics—the study of how traits (or genes) are passed from one generation to the next—which was the main focus of these early geneticists. There is a parallel here between the objective of transmission genetics and what Weismann was up to as he tried to understand and explain how traits can move between generations.

Early twentieth-century transmission genetics, then, was keenly focused on the problem of heredity, and this problem became framed in terms of the movement of traits between generations and the materials involved at the sub-cellular (e.g., chromosomal) level. Although many of these scientists, including Morgan, trained within embryology and cytology and botany, their research programs within genetics were largely devoid of thinking about heredity in terms of development of the embryo. But just because they eschewed embryology and (largely) cytology does not mean that they forgot their training in these fields.

Here is where Wilson's book comes back into play. Wilson's editions of *The Cell in Development and Inheritance* were standard reading for practitioners in the fledgling field of transmission genetics. And in this book, Wilson gave these geneticists in the 1900s through 1920s the perfect means of simplifying their understanding of an organism's heredity: the Weismann Barrier. The incipient field of transmission genetics came to rely fundamentally on the Weismann Barrier to presuppose the continuity of the hereditary material contained within the germ cells between generations and the non-inheritance of acquired characteristics. By assuming that germ cells were something separate and distinct, that they had their own separate cell lineage, and that this lineage was continuous within organisms and between generations, as Wilson had indicated that Weismann claimed, transmission geneticists could simply point to this cell lineage and say these were the cells that held the keys to heredity. Because they were focused on the chromosomes and not even the germ cells, these early geneticists did not need to concern themselves with the origins and traits of the cells, so they didn't.

Thus, the Weismann Barrier and the notion that the germline is a separate and distinct cell lineage to which somatic cells cannot contribute became embedded in the thinking of transmission geneticists during the first decades of the twentieth century. This thinking only became more deeply rooted over time. In 1957, American paleontologist George Gaylord

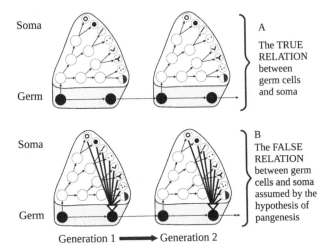

FIGURE 1.5 | Recreation of figure 12.1 from Simpson et al., *Life: An Introduction to Biology*. The figure is labeled "The one-way relation between germ cells and the differentiated cells of the soma." Note how the figure is captioned: the diagram showing the Weismann Barrier (*A*) is called the "true" relation between germ cells and soma, while the figure showing possible feedback from somatic cells to germ cells (*B*) is made equivalent to Darwin's theory of pangenesis and labeled "false." Created with BioRender.com.

Simpson, who had strong ties to the genetics community throughout the middle of the twentieth century, along with two colleagues, wrote a biology textbook, *Life: An Introduction to Biology*, which became popular among US audiences. Simpson and colleagues provide a detailed discussion of the history of the chromosome theory of heredity—the basis of transmission genetics and the subject for which Thomas Hunt Morgan won his Nobel Prize in 1933. In a section of this discussion titled, "Weismann: The One-Way Relationship between Germ Cells and Soma," of which figure 1.5 is part, Simpson and colleagues discuss the Weisman Barrier.

Here we can see a slight modification to Wilson's figure from 1896, with an even more definitive declaration about the relationship between germ cells and somatic cells. In the text accompanying this figure, Simpson and colleagues tell the reader,

> From which type of cell is the egg or sperm derived? It can-
> not be a descendant of all these differentiated types: a cell is a
> descendent of a single cell. Weismann provided the answer to
> this problem by pointing out that the germ cells of each gener-
> ation were direct descendants through a lineage of unspecial-
> ized cells from the germ cells of the previous generation. That
> is, the specialized <u>body cells of each generation are related to</u>
> <u>germ cells in a one-way fashion: they are derived from germ</u>
> <u>cells but do not give rise to them</u>. This insight is a death blow
> to pangenesis in all its forms.[36]

Simpson's figure and text give us an indication of just how important the Weismann Barrier had become for geneticists by the mid-twentieth century. To them, it was the cornerstone of the chromosome theory of heredity, which, in turn, was the major breakthrough of genetics in the early twentieth century. One of the most famous evolutionary biologists of that century, US-based German scientist Ernst Mayr, reinforced this thinking further. With Simpson, Mayr gained fame for leading a movement in the mid-twentieth century called the Modern Synthesis, which conceptually brought together Darwinian evolution, transmission genetics, and systematics (the branch of biology that deals with classification of species).[37] As Mayr approached the final decades of his career, he took up writing history. In his 1982 book, *The Growth of Biological Thought:*

Diversity, Evolution, and Inheritance, he attempted to catalog the conceptual history behind the science of his work:

> This observation [of the early segregation of the germline] led Weismann in 1885 to his theory of the "continuity of the germ plasm," which states that the "germ track" is separate from the body (soma) track from the very beginning, and thus nothing that happens to the soma can be communicated to the germ cells and their nuclei. We now know that Weismann's basic idea—a complete separation of the germ-plasm from its expression in the phenotype of the body—was absolutely correct. His intuition to postulate such a separation was faultless.[38]

Mayr appears to conflate Weismann's notion of germ plasm continuity with the idea that germ cells are continuous and are separated from the somatic cells "from the very beginning." Mayr is thus furthering what Wilson and Simpson and colleagues had made explicit: that the germ cells were a separate cell lineage from the somatic cells, that somatic cells could never become germ cells, and that it was conceiving of the continuity of the germ cell lineage, or germline, that enabled the field of genetics to advance theories of heredity.

These scientists—Wilson, Simpson, and Mayr—were preeminent and esteemed members of their respective fields of cell biology, paleontology, and evolutionary biology. Bolstered by their influence, the validity and importance of the Weismann Barrier became entrenched within biological thought, and the Weismann Barrier became a keystone of the history of genetics.[39]

CONCLUSION

Throughout this chapter, we have seen how scientists thought about the origins and nature of germ cells, beginning with Leeuwenhoek's discovery of spermatozoa in 1677 and going on through the establishment of the Weismann Barrier as a cornerstone of the geneticists' view of heredity in the twentieth century. We have also seen how the Weismann Barrier was not established from biological facts *per se*—it was established by misinterpreting the theoretical work of its namesake, August Weismann. But this history is only part of the story of how the Weismann Barrier became entrenched within many scientists' thinking and remained a pervasive assumption within considerations of germline. So far, I have traced how the Weismann Barrier was absorbed by the fledgling fields of transmission genetics and propagated in the context of thinking about heredity. The other part of the story involves understanding how the Weismann Barrier—a concept that indicates a relationship between cells—became established within embryology and how the other two assumptions that underlie thinking about germline regeneration that I laid out in the introduction arose. For this part of the story, we need to move to the next chapter.

2   Backgrounding Conflicts within Germ Cell Science

We have now seen how transmission geneticists absorbed the Weismann Barrier as a main tenet of their thinking during the early twentieth century: Germ cells can become somatic cells, but somatic cells can never become germ cells. Embracing this assumption within transmission genetics was not based on careful investigations of the origins and nature of germ cells; rather it was a convenient way of doing exactly the opposite — setting the details of cells and development to the side. In effect, these early geneticists treated what should have been an empirical claim about the relationship between germ cells and somatic cells as a normative claim. An empirical claim can be determined to be true or false via evidence; whereas a normative claim defines the way in which something ought to operate.

By treating the Weismann Barrier as a normative claim, early transmission geneticists could simply point to the germline and say that germ cells held the keys to heredity because they operated in the way that the Weismann Barrier dictated. But the problems of the origins of germ cells and germinal continuity didn't go away simply because transmission geneticists refused to take them into account. Instead, they became the purview of embryologists, who were not content simply to accept the relationship between germ cells and somatic cells that the Weismann Barrier dictated.

As we'll see, for some embryologists, the entire premise of the Weismann Barrier was absurd—there was no fundamental difference between germ cells and somatic cells; it was a matter of differentiation as with any other kinds of cells. For others, the distinction was the cornerstone of their understanding of development and organismal biology.[1]

In the end, the diversity of perspectives coalesced around a more unified embrace of the Weismann Barrier during the middle of the twentieth century. This shift was not a simple one. Disagreements about the nature and origins of germ cells, like most problems within the sciences, did not disappear overnight. Rather, as the century wore on, the assumptions that drove earlier investigators into conflict gradually faded into the background. These assumptions have lingered in the background of scientific research and thinking about germ cells, germline, and germline regeneration and still shape our understanding of these things.

THERE IS NOTHING SPECIAL ABOUT GERM CELLS

In 1931, American embryologist Florence Heys provided an extensive review of the search for the origins of germ cells and their fates in which she laid out exactly how disparate views were at this time.[2] According to her analysis, scientific thinking about germ cells in the first few decades of the twentieth century existed along a spectrum. At one end were those who didn't think that there was anything special about germ cells—they were simply another type of cell—therefore, there was no Weismann Barrier, and no continuity of germ cells within organisms or between generations. At the other end were those who saw germ cells as unique, separate, and distinct from somatic cells, as a continuous lineage, and the relationship between germ cells and somatic cells as upholding the Weismann Barrier. Others held some combination of these views.

American embryologist and zoologist, George T. Hargitt, was one of the researchers who thought there was nothing special about germ cells. Throughout his research on germ cells, Hargitt investigated a wide array of species and was one of just a few scientists who studied germ cells in both vertebrates and invertebrates. From his earliest research concerning germ cells, Hargitt became convinced that methodological assumptions about identifying germ cells had tainted the conclusions that many of his contemporaries had reached.

Between 1913 and 1919, Hargitt produced a series of research articles that tracked the origins of germ cells within a series of hydrozoan species—small, marine invertebrates related to jellyfish and coral—not coincidentally the same group of animals that led Weismann to his germ plasm theory.[3] Hargitt wanted to know where and when the germ cells originated within these different hydrozoans, and so he made careful studies of their embryos and larvae. From these years of staring at hydrozoan embryos, larvae, and adults, Hargitt determined that the germ cells differentiate at the time just before sexual maturity, and that they appear to arise from functional somatic cells. He concluded about hydrozoans that "there is no definite migration of germ cells and no germ-track; there is no invisible germ plasm in the body cells."[4] Hargitt's investigations of hydrozoans led him to believe that there was nothing unique about germ cells—they were simply differentiated somatic cells—and that there was no continuity of a germ cell lineage or an invisible and unobservable germ plasm.

Following the publications of his results, several researchers intimated that Hargitt's conclusions about germ cells and germ plasm might well apply to hydrozoans but that such evidence could not be used to dismiss this idea within vertebrates.[5] In turn, Hargitt extended his studies to vertebrates. In 1924, he published his findings on the origins of germ cells in the salamander, *Diemyctylus viridescens*, also known as the crimson-spotted newt.[6] Due to the differences between the life cycles of hydrozoans and vertebrates, when examining *Diemyctylus*, Hargitt divided the problem of the origin of germ cells in vertebrates into two phases. The first phase was concerned with the earliest appearance and source of primordial germ cells during development and their subsequent fate. The second phase dealt with the relationship in adults between these primordial germ cells and subsequent germ cells. Hargitt focused his attention on the second phase in *Diemyctylus*. He collected and observed tissues of the testes and noticed that germ cells were always found in the supportive tissue surrounding the structures of the testes, called *stroma* (a somatic structure). The consistent location of these germ cells implied to Hargitt that they were differentiated from the surrounding somatic tissue. Because he could not find evidence that germ cells that had segregated early during development led to these germ cells that he witnessed in the gonads, he believed that germ cells arose from the somatic cells of the gonads. These observations led Hargitt to conclude that his findings from hydrozoans were true for vertebrates (at least *Diemyctylus*).

Hargitt continued his studies of germ cell origins in vertebrates by investigating the earliest appearance and source of primordial germ cells during development and their subsequent fate in rats.[7] Hargitt studied embryos in stages from just before germ layer formation (the germ layers are three tissues that give rise to all subsequent parts of the organism) to the beginning of gonad formation. He carefully cultivated embryos, fixed them in solution to preserve them, cut them into sections, and then applied various stains to the sections.[8] He began by observing the later-stage embryos in which the gonads had begun to form and germ cells could be definitively identified, and worked his way backward, reasoning that this approach would allow him to trace any germ cells that contributed to the population in and around the presumptive gonad. From his observations, Hargitt concluded that the germ cells do not appear early during development, and that there is no germ track in the rat. Thus, through his studies of rats, Hargitt again confirmed the conclusions that he had reached about the origins and nature of germ cells from his early work on hydrozoans: that there was nothing unique about germ cells—they were simply differentiated somatic cells—and that there was no continuity of a germ cell lineage or an invisible and unobservable germ plasm.

There were two underlying and intertwined methodological assumptions that Hargitt observed in previous scientists' work on the origins of germ cells. The first was that scientists could accurately identify primordial germ cells within the developing embryo. Primordial germ cells are the first type of germ cell in the body, which are supposed to differentiate well before the gonads form. In other words, scientists thought they knew what primordial germ cells looked like. But Hargitt tells his reader that "the fact is there is no single character which will distinguish germ cells from other cells, nor is there any combination of characters which may be used indiscriminately for such a test."[9] He meant that none of the methods or criteria that scientists had developed to identify primordial germ cells was specific to these cells, and thus they were not reliably accurate. Let's take a closer look at these identification methods.

From the late nineteenth century through well into the twentieth century, scientists relied on two means of identifying primordial germ cells: (1) morphological criteria, and (2) selective staining criteria. There were a lot of morphological criteria that were supposed to distinguish germ cells from somatic cells: for example, the presence of granules in the cytoplasm, the appearance of the mitochondria (a subcellular structure), the large size of the cell, the round shape of the cell, and finally, well-defined cellular membranes.[10] Hargitt believed that each of these morphological criteria was problematic. For instance, the presence of a specific type of mitochondria in germ cells was definitively disproved just a few years after it was identified in 1910.[11] Meanwhile, granules were not found in the cytoplasm of germ cells in all species. And, while germ cells like ova are certainly large, round, and have well-defined cell membranes, these characteristics apply to other cells as well. Most cells tend to round out when approaching mitosis (cell division), and they acquire a more well-defined cell membrane. As for the size of the cell, Hargitt tells his reader that "it seems pretty clear that practically every cell of large size in any part of the embryo has been called a germ cell."[12]

Selective staining criteria refers to applying chemicals to cells and or embryos. These chemicals are supposed to color or highlight specific subcellular parts, which allow scientists to differentiate between cell types. Researchers developed quite a few selective staining criteria for germ cells during the late nineteenth and early twentieth centuries. For example, in 1912, Russian physician and biologist W. Rubaschkin described staining his mouse embryos with a mixture of chemicals called azure II (which gives a blue tint) and eosin (which gives a red or pinkish tint).[13] According to Rubaschkin, primordial germ cells have a different reaction to this combination of cell stains than all other cells, so primordial germ cells can be identified by their red-tinted nuclei and pale blue cytoplasms. In 1923, American embryologist Cleveland Simkins described how inaccurate Rubaschkin's method was, highlighting how cell cycle phase heavily affects the way in which cells take up stains: many different types of cells that are preparing for mitosis will absorb eosin and present with a red nucleus.[14] In 1925, Hargitt had this to say about selective staining criteria: "The staining reaction of germ cells is sometimes stated to be different from other cells, but this is a very uncertain test, for the germ cells even in the same section do not by any means stain constantly. . . . All staining reactions depend upon the physical and chemical conditions, and these are correlated with the metabolic state of the cells; we have no right to expect any positive or constant difference in staining reactions of germ cells and other cells."[15]

Thus, Hargitt found many of the selective staining criteria that earlier scientists had developed to be dubious and inaccurate, and he maintained that they could not be used to definitively identify primordial germ cells. In Hargitt's estimation, then, the morphological and selective staining criteria that scientists used to identify primordial germ cells were so nonspecific that it led them to assume the presence of germ cells during the pre-gonadal stages of development. In other words, identifications of germ cells before the gonads formed were based on faulty assumptions.

The second methodological assumption that Hargitt saw is interconnected with the first: many of his predecessors and contemporaries believed that they could identify primordial germ cells without extensive investigations of development. If, Hargitt reasoned, someone wanted to make a claim about where and when germ cells originate before the formation of the gonads, they would need to trace the lineages of the germ cells that could be definitively identified within the gonads, because there was no reliable means of identifying germ cells before this stage. In other words, not many studies of germ cell origins were exhaustive cell lineage studies, and this was especially true for vertebrates. You can see how these two assumptions build on each other—without a means of accurately identifying primordial germ cells before the formation of the gonads, scientists could not just point to cells in early-stage embryos and call them germ cells. They would need to trace the cell lineages from the germ cells within the gonads backward through development to identify primordial germ cells definitively, and this was not a common practice.

Hargitt found these two methodological assumptions to be particularly problematic because they led many of his contemporaries to claim that germ cells and germ plasm were continuous—a proposition that Hargitt found no evidence to support.

THE WEISMANN BARRIER REIGNS

At the opposite end of Heys's spectrum of thinking about germ cells in the early twentieth century is American invertebrate zoologist Robert Hegner, whose driving question was: What makes a germ cell a germ cell?

Hegner's first attempt at addressing this question came in 1908, when he spent several months experimenting on leaf beetles.[16] He identified a dark, granular substance in the posterior end of freshly fertilized embryos that he dubbed the "pole-disc." Hegner watched as these embryos developed and tracked how the cells divided and migrated around the embryo. He noticed that some cells migrated through the pole-disc, giving them a characteristic dark color and the appearance of granules within their cytoplasm (recall my prior discussion of morphological identification criteria), and they came to reside as a clump at the posterior end of the embryo. He called these "pole cells." Following the development of the pole cells further, Hegner determined that they became the primordial germ cells (the first type of germ cell in the body). He noticed that during cell migration within these early embryos, other cells came into contact with the pole-disc, but did not become pole cells—they didn't take on the dark appearance or have granules within their cytoplasm. Hegner reasoned that the pole-disc granules bore some responsibility for the formation of the primordial germ cells, and introduced the term "germ cell determinants" to describe these granules.

Over a series of experiments, Hegner tested his reasoning. He tried tying off the posterior end of freshly laid eggs with silk thread to remove the pole-disc; it didn't work—the eggs burst. He then came up with a way to puncture freshly laid eggs on the posterior end such that they would extrude the pole-disc material (and often a lot else). The results of these early experiments were not definitive, but they gave a good indication that when the pole-disc was removed, the primordial germ cells would not form. Over the next few years, Hegner continued to refine his experiments on removing the pole-disc from freshly laid insect eggs.[17] He gradually got more definitive results until 1911, when he declared that, "the pole-disc granules *are* necessary for the formation of germ cells, and that they are really 'germ cell determinants.'"[18]

The contribution of cytoplasmic granules to primordial germ cells was well documented before Hegner's experiments, if not underappreciated. In 1863, Weismann noticed the contribution of cytoplasmic granules to what he called *"Polzellen"* (German for "pole cells") during his observations of two species of insect, but did not yet recognize that the pole cells became the primordial germ cells.[19] In 1865, the Russian-French zoologist (and later a groundbreaking immunologist), Élie Metschnikoff, as well as the German botanist and illustrator Rudolf Leuckart, independently verified that the pole cells became primordial germ cells.[20] In the intervening years between Metschnikoff, Leuckart, and Hegner, a whole host of other scientists recognized the presence of granules within pole cells and also noted that primordial germ cells tended to contain a lot of dense, yolk-like globules in their cytoplasm.

What Hegner proposed in 1908, and later confirmed, was that the cytoplasmic granules within the pole cells, what he called "germ cell determinants," was "the material which fixes the character of the cells."[21] Hegner thus believed that germ cell determinants were the substance that made a germ cell a germ cell. In 1914, Hegner switched from calling the cytoplasmic material "germ cell determinant" to "keimbahn-determinant."[22] Recall from the previous chapter that keimbahn was the German term Weismann coined, which became "germ-track." In renaming the cytoplasmic materials that Hegner thought gave germ cells their identities, he was explicitly referencing Weismann and his germ plasm theory. But, he came to some very un-Weismann-like conclusions about the nature of germ cells.

Weismann had thought of the germ-track as the lineage of cells that leads from the single cell of the fertilized egg to the emergence of the first germ cell and believed that it was composed of somatic cells that carried a full complement of unaltered germ plasm. For Weismann, germ cells were a specialized type of somatic cells and were especially important for their unique roles in heredity; they were not special because of a fundamental difference in their structure or their substance. After all, unaltered germ plasm was also present in cells in areas of the body that were susceptible to regeneration. Hegner's thinking about the *keimbahn* aligned with Weismann's only insofar as he, too, envisioned a lineage of cells. However, for Hegner, the cells in this lineage were *all* germ cells, because they contained *keimbahn*-determinants—a cytoplasmic material that gave germ cells their identities.

In his 1914 book called *The Germ-Cell Cycle in Animals*, which was a massive review of research on germ cells, almost entirely based on studies of invertebrates, Hegner evaluated the two theories of continuity that had been ascribed to germ cells: germinal continuity (germ plasm continuity) and morphological continuity (cellular continuity). Regarding germinal continuity, Hegner agreed with Weismann—there was a continuity of germ plasm. But, relying on his own studies and evidence from other invertebrate studies, Hegner sought to expand on Weismann's understanding of the germ plasm. In considering germ plasm continuity, Hegner wrote, "If, then, we accept germinal continuity as a fact and consider the germ-plasm to be a substance that is not contaminated by the body in which it lies, but remains inviolate generation after generation, we should next inquire as to the nature of this substance."[23] Although Hegner mused that the relationship between his *keimbahn*-determinants and the germ plasm was "not yet definitely known,"[24] given his experimental evidence, he concluded that the *keimbahn*-determinants located within the cytoplasm were made of this substance. After all, Hegner reasoned, if Weismann's germ plasm theory simply required a substance to be reserved for germ cell formation, the *keimbahn*-determinants certainly fit this description. I should mention that while Hegner did have reasonable evidence to believe that his *keimbahn*-determinants in some way were connected with the identity of primordial germ cells, he also had evidence that they (1) were not the same in all species in which they were found and (2) were not found throughout all species examined. He didn't consider either of these points problematic for his conclusions.

When reviewing morphological continuity, Hegner began by telling his reader that "No case of a complete morphological continuity of germ cells has ever been described."[25] Despite this admission, Hegner assumed that it existed and that germ cells were continuous—he contended that the evidence just hadn't caught up to fully justifying this view. We can see how Hegner could come to such a conclusion, given his assumptions of germ plasm continuity and the *keimbahn*-determinants as in some way connected with Weismann's germ plasm. For him, primordial germ cells were defined by the presence of *keimbahn*-determinants, which were somehow connected to germ plasm, and germ plasm was continuous. Therefore, germ cells must be continuous.

Hegner thus connected his *keimbahn*-determinants with Weismann's germ plasm, creating a conceptual shift in the material from the nucleus to the cytoplasm. In doing so, Hegner inadvertently conflated Weismann's view of germ plasm continuity with a view of germ cell continuity. He also shifted the role and function of the germ plasm from the material that could account for heredity and all cellular differentiation to a material that defined the identity of germ cells. These views proved influential with subsequent scientists.

In 1934, French biologist Louis Bounoure published the results of his studies of the origins of germ cells in the frog, *Rana pipiens*. While observing developing frog embryos, Bounoure noticed the presence of cytoplasmic granules that segregated into the primordial germ cells. Referencing Hegner's earlier work, Bounoure called these granules *"cytoplasme germinale,"* which translates to "cytoplasmic germ plasm," or simply "germ plasm."[26] Five years later, Bounoure published a massive review of germ cell research in a book titled, *L'origine des cellules reproductrices et le problème de la lignée germinale.*[27] In this text, Bounoure suggests that the germ plasm that he had described in the cytoplasm of frog eggs years earlier was responsible for determining the germ cell lineage (Hegner's *keimbahn*).

Bounoure's book had a large impact on embryologists and, later, developmental biologists interested in germ cells. It was the first extensive review of germ cell research since Hegner's 1914 text. Other reviews appeared within this interval, such as Heys's 1931 article, but they were far less extensive than Bounoure's book and tended to shy away from theoretical claims. Bounoure had no such compunction, and he declared both that the continuity of the germ plasm was a general law of organismal reproduction and that the cellular continuity and early differentiation of the germ cells could not be doubted.[28] Bounoure saw the germ plasm—the cytoplasmic substance that Hegner had connected with germ cell identity—as the material that defined the germ cells and made them continuous within and between generations.

CONFLICTING ASSUMPTIONS

Looking at the research programs of George Hargitt and Robert Hegner and the conclusions they reached gives us good insight into where the conflicts about the origins and nature of germ cells and germinal continuity lay in the first few decades of the 20th century, and what assumptions underpinned these conflicts. Let's briefly look at these assumptions.

First, as Hargitt and others pointed out, there were methodological assumptions around identifying germ cells. Hargitt laid out in detail the issues that surrounded the morphological and selective staining criteria that scientists used to identify primordial germ cells. He showed how problems with identifying these earliest germ cells led to problems with reaching conclusions about the early origin or continuity of germ cells. This in turn made scientists who used these methods without considered diligence vulnerable to introducing a lot of error into their understanding of the origin of germ cells. Hargitt was not the only scientist at the time who was concerned with the methods of identifying germ cells and the conclusions to which these methods led. However, these concerns were not pervasive in the literature.

Second, as we saw in the case of Hegner and Bounoure, a lot of assumptions surrounded the problem of continuity (of germ plasm and of germ cells) and the nature of the germ plasm.

Hegner's movement of the germ plasm from the nucleus to the cytoplasm allowed him to call the germ plasm continuous, and we saw how his reasoning about the role of the germ plasm led him to conclude that germ cells were also continuous. Hegner did not have evidence to support a continuous germ cell lineage, and as I pointed out, there were some issues with conceiving of a continuous germ plasm. As Hargitt noted in 1944, Hegner's proposal to shift the germ plasm to the cytoplasm and change its role from heredity to germ cell fate was a major leap in understanding the germ plasm. Hargitt wrote, "What a strange reversal: from the view that germ cells differ from tissue cells because of diverse nuclear composition to the view that all nuclei contain complete hereditary material of the species, and the differences between cells are due to the kind of cytoplasm in which the nuclei are located!"[29]

More than two decades later, Bounoure approached continuity with an even more dogmatic style. While Hegner couched his conclusions in qualifying language, like noting that the relationship between Weismann's nuclear germ plasm and his own cytoplasmic *keimbahn*-determinants were "not yet definitely known," Bounoure made no such concessions.[30] As the prominent British evolutionary biologist Julian Huxley, who along with George Gaylord Simpson and Ernst Mayr founded the Modern Synthesis, put it in his review of Bounoure's book, "For him [Bounoure], germinal continuity is not a problem to be investigated, but a doctrine, or rather a dogma, to be proclaimed and supported."[31] In other words, Bounoure assumed that there was germ cell and germ plasm continuity, even when he did not have direct evidence to support such an assumption.

We could ask why Hegner and Bounoure were so drawn to the idea of continuity that they would make strong claims about it without definitive evidence, or why many investigators ignored Hargitt's call to question their methods of identifying Primordial Germ Cells. Looking at the state of germ cell research in 1948, embryologists John Normal Berrill and Chien-Kang Liu give us a good answer to these questions: "The weight of authority . . . of the Weismann-Nussbaum combination [germ cell continuity] convinced many later workers of the existence of facts they could not observe, and much subsequent argument has arisen over the identity of so-called primordial germ cells and the existence of a germ-track in developmental stages younger than those in which germ cells can be safely recognized."[32] Some scientists, then, may have been so tied to the *idea* of continuity that they were willing to overlook both the lack of evidence to support it and the potential problems with the evidence that they did have. This way of thinking was not tied solely to Hegner and Bounoure; it was widespread within studies of germ cells at the time.

THE FALSE VITALITY OF THE WALKING DEAD?
COALESCING AROUND GERM PLASM AND CONTINUITY

In 1948, Berrill and Liu evaluated the evidence and thinking behind research on germ cell origins and claims of germ cell and germ plasm continuity. They noted how investigators like Hargitt and Cleveland Simkins made strong and condemnatory attacks against the notion of the continuity of germ cells and germ plasm, but despite these attacks, the idea always regained strength. This led Berrill and Liu to ponder whether "this vitality [of the continuity of germ cells and germ plasm] may of course flow from truth itself incorporated in the germ-plasm concept, or it may be a false vitality akin to the walking-dead habits of Dracula."[33] It's worth taking Berrill and Liu's point seriously.

Less than thirty years after Berrill and Liu's paper was published, developmental studies of the origins and nature of germ cells had coalesced around the conceptual work of Hegner that Bounoure had dogmatized. In 1976, American biologist Edward Mitchell Eddy published a massive review article, titled "Germ plasm and the differentiation of the germ cell line," which has proven influential, as evidenced by its citation record.[34] In this paper, Eddy laid out the state of thinking about germ cells. By this point, germ plasm was considered to be "a substance present in the cytoplasm of gametes, which is segregated into specific cells during blastulation and determines that those cells shall become the progenitors of the germ cell line during subsequent development."[35] Eddy made reference to Weismann's concept of germ plasm continuity, but indicated that this concept had shifted to something else entirely: "Although when Weismann wrote of the continuity of germ plasm he was referring to genetic material of the nucleoplasm, this account forms the basis of the present germ plasm hypothesis."[36] The germ plasm hypothesis, indicating that the germ cell line is determined by an agent present in mature germs cells, suggests that the agent contains stored information capable of influencing the formation of primordial germ cells."[37] By 1976, then, Hegner's vision of the germ plasm and the conflation of germ-plasm and germ cell continuity that it had initiated, had become standard within germ cell research. The germ plasm was now understood as a cytoplasmic material that determined the germ cell lineage. Germ plasm *and* the germ cells were continuous. Eddy's portrayal of thinking about germ cells reflected a much broader acceptance of the germ cells and germ plasm along these lines. Such thinking about the origins and nature of germ cells has continued in this vein into quite a bit of current germ cell research.

The widespread acceptance of the reenvisioned germ plasm, and of germ plasm and germ cell continuity, could lead us to think that the answer to Berrill and Liu's question comes down on the side of "truth itself incorporated in the germplasm concept." But is that really the case? The assumptions I have dissected from early twentieth-century studies of germ cells were not resolved by the time Eddy published his review in 1976. Instead, they had faded into the background. Lurking. Waiting. Rather like Dracula. My task is to bring them back into the light and, following the Dracula analogy, see if they turn to ash. The question now becomes—do these assumptions give us reason to see the Weismann Barrier and associated thinking about germ cell continuity, as having the false vitality of the walking dead? To answer this question, we need to turn to more current research and see whether all of the historical assumptions that underpin germline regeneration that I've highlighted in the last two chapters are reasonable. And, if not, what can be done to overcome them.

3   Challenging Assumptions in Germline Science

We are left with three empirical claims about the Weismann Barrier, continuity, and germ cell identification. Although these empirical claims are drawn from historical research, they are not often explicitly addressed by current germline research, and their history shows that they are unresolved. We can therefore think of these claims drawn from history as assumptions—as things that scientists often take for granted within their research programs.

This brings me back to the driving question of this book: *How does the germline regenerate?* In order to answer this question, I need to challenge the assumptions that I have uncovered in the history of thinking about germ cells, which underlie a lot of current thinking about germline regeneration. Challenging these three assumptions means prodding the reasoning and evidence that support them. Such prodding has potential repercussions for what we think we know about germ cells, including germline regeneration.

The common framing of germline regeneration among scientists involves the following reasoning: Only germ cells can

regenerate lost or damaged germ cells. To support their statement that the germline can only regenerate from germ cells,
scientists must (1) be able to accurately and reliably distinguish
between germ cells and somatic cells; (2) be able to say definitively that the germ cell lineage is continuous, meaning that
there is an unbroken line of descent among germ cells within
an organism; and (3) be able to state that under no circumstances can somatic cells become germ cells. You can see how
these assumptions are intertwined and build on one another.
Scientists must be able to accurately distinguish germ cells
from somatic cells in order to make a claim about germ cell
continuity, and germ cells must maintain a continuous and
unbroken line of descent throughout development in order
for the Weismann Barrier to hold. My task now is to determine whether these assumptions that arise from the history of
thinking about germ cells are valid, or whether they are invalid
and thus give our understanding of germ cells, and associated
thinking about germline regeneration, the false vitality of the
walking dead.[1]

IDENTIFYING GERM CELLS

As we have seen, questioning the methods that scientists use
to identify germ cells extends back well over a century. George
Hargitt and a handful of other scientists explained the problems inherent in morphological criteria and selective staining
criteria in the first decades of the twentieth century. Hargitt
drew attention to and questioned these methods to emphasize
the fact that scientists can only determine the nature of the

relationship between germ cells and somatic cells if they can reliably and accurately identify and distinguish between these types of cells. He pointed out that methodological problems with identifying germ cells, especially primordial germ cells, can lead scientists to construct a false narrative about when and where germ cells originate and how they may relate to each other over an organism's life history. Misidentifying cells in the embryo can lead to a false sense of continuity between germ cells and the relationship between germ cells and somatic cells. There were certainly grounds to object to the methods of germ cell identification during the early twentieth century, especially for primordial germ cells, and many of these objections still stand.

By the 1990s, scientists had more tools available to help them identify germ cells, thanks to advances in understanding heredity and development through genetics. Many had begun to use gene expression, in addition to morphological and selective staining criteria, to identify the germ cells. Gene expression is the process by which genes synthesize their end products, like proteins or non-coding RNAs, and indicates when and where a particular gene is active within a cell or organism. Scientists have a variety of methods for determining and visualizing when and where genes are expressed. We tend to think of genetic information as being far more accurate than these other criteria, but there are limits to this method of identifying germ cells. Over the past thirty years, scientists have uncovered a whole suite of genes whose expression in cells can be and has been used to identify germ cells, such as *Piwi, Nanos, Vasa, Pumilio,* and *Tudor.*[2] But, just as with the morphological criteria and selective staining techniques discussed earlier, these genes are usually not expressed solely in the germ cells. In fact, this group of genes is also used to identify what scientists call *multipotent stem cells*. These are cells that have the capacity to self-renew by dividing and to develop into multiple, specialized somatic cell types present within tissues or organs. Because this suite of genes is not specific to germ cells, there are grounds to question whether using such gene expression methods alone is sufficient to distinguish between germ cells and somatic cells. Questioning the accuracy of this method, however, is not common within current science, just as questioning morphological or selective staining criteria wasn't common in Hargitt's time.

One way to improve the accuracy of using genetic methods for identifying germ cells is to do what Hargitt suggested in 1925—trace the entire germ cell lineage. Whereas those intrepid biologists in the late nineteenth century spent hundreds of hours collecting, preserving, staining, and observing embryos, recent genetic techniques make tracing cell lineages more straightforward. Cell-lineage tracing has changed since Whitman and his colleagues introduced it more than a century ago, but the logic behind it remains the same—trace the relationships and ancestry of cells through development. Tracing cells in the late nineteenth century relied primarily on direct observations of cell divisions under the microscope, combined with observations of preserved embryos treated with selective stains. Now scientists can use a combination of selective cell marking, genetic techniques, and microscopy to pinpoint cells of interest during embryonic development and see where they go and what cells they give rise to. However, even with the advent of new techniques to trace cell lineages, the process is still extremely time and resource intensive, and most studies of germ cells do not incorporate a cell-lineage tracing element.

Identifying germ cells, then, is still complicated. Scientists still rely on a combination of morphological and selective staining criteria, in addition to gene expression criteria and other techniques. If scientists have a good understanding of the entire germ cell lineage within an organism, which they can gain today with lineage-tracing techniques, then using these identification techniques can be fairly accurate. But doing so consumes a lot of time and resources, and these techniques are not used by the majority of scientists who study germ cells. The methodological assumptions that Hargitt pointed to in the early twentieth century, then, are still alive within current studies of germ cells. Therefore, any confident assertion that germ cells and somatic cells maintain separate and distinct lineages should be treated with caution.

CONTINUITY

Earlier we saw how scientists in the late nineteenth and early twentieth centuries thought of continuity of germ plasm (Weismann), continuity of germ cells (Wilson, Simpson), or a mashup of both (Hegner, Bounoure following revisions to the meaning of germ plasm as a cytoplasmic identifier of germ cells, and Mayr) as they tried to wrangle an understanding of the origins and nature of germ cells. The problem of continuity has not gone away in the intervening century plus of research.

In current science, many investigators consider the germ cell lineage to be continuous, but what continuity means often is not made explicit, because it is no longer the focus of intensive debates, as it was with our historical actors. Just because the assumption of continuity has been backgrounded within current science does not mean that we should consider the matter settled. Challenging this historical assumption, then, requires us first to sort out what continuity of germ cells or germline means within current research, then we can ask whether or not the ideas hold.[3]

In the works of Hegner and Bounoure, the concept of germ continuity conflated material continuity of the germ plasm and cellular continuity. Although most current concepts have escaped this conflation, we can still use the works of Hegner and Bounoure as a springboard into more modern works by categorizing thinking about continuity into two areas: (1) as a continuity of germ plasm, and (2) as a continuity of germ cells as a cell lineage.

Let's begin by thinking about the continuity of germ plasm. Thanks to the work of Hegner and Bounoure, germ plasm in modern germ cell biology refers to a material clumped within the cytoplasm of germ cells, which we now know to be made of proteins and mRNAs. Germ plasm is supposed to play a role in defining germ cells and be passed on from parents to their progeny, resulting in a material that is both contained in all germ cells within an organism and passes between generations (via presence in oocytes). Making this claim, some scientists following Hegner and Bounoure called germ plasm "continuous." While this notion of germ plasm continuity is not as widespread today as it was in the early-to-mid-twentieth century, it did not die with Bounoure. For example, in 1974, biologists Jeffrey B. Kerr and Keith E. Dixon described germ plasm continuity as follows, "Implicit in the idea of a separate germ line is the concept that germ plasm is continuous from generation to generation, and also therefore within a genera-tion. . . . For continuity between generations, it is only neces-sary for the female germ cells to carry the germ plasm."[4] More recently, a series of scientists have called the germ plasm the basis of germ cell continuity.[5] Some current scientists, then, adhere to the notion of continuity that Hegner and Bounoure started, and on that basis, they conclude that there is a contin-uous cell lineage.

The notion of a continuous germ plasm faces two serious challenges. First, while germ plasm plays a role in determining the primordial germ cells in some species like fruit flies (*Dro-sophila melanogaster*), it has not been found in all organisms. You may recall that Hegner mentioned this in his 1914 book. If germ plasm is the basis of germ cell continuity, then we would anticipate that the germ cells of all organisms contain germ plasm, which they do not. Second, in organisms that have germ plasm, there is no evidence that the germ plasm persists unal-tered throughout the germ cell lineage or across generations. Given these two challenges, it is problematic to define conti-nuity on the basis of germ plasm because the definition cannot be applied to all organisms or even definitively said to occur in organisms that use germ plasm to differentiate their germ cells.

The more prevalent notion of continuity within the scien-tific literature is associated with the idea of germ cells as a continuous cell lineage, also known as the germline. The *Oxford English Dictionary* defines germline as "a series of germ cells descended from earlier cells in the series, regarded as continuing through successive generations of an organism."[6] Continuity, here, means that there is an unbroken line of descent among germ cells within an organism and between generations. This corresponds to the thinking that Jaeger and Nussbaum laid out in the nineteenth century. This way of viewing continuity does not rely on the presence and activity of germ plasm. However, how a germ cell lineage can remain continuous from one generation to the next is far from clear. This lack of clarity concerns two different, albeit related, problems: continuity of a germ cell lineage *between generations* and *within an organism*. In order to understand the constraints on both of these ways of conceiving of germ cells, we need to look first at how germ cells originate.

Germ cells originate during development through a process of differentiation (often referred to as *specification*) that gives rise to the primordial germ cells. This differentiation process occurs in slightly different ways across species, but there is currently a scientific consensus that specification happens in one of two ways: either maternal inheritance (also called *preformation*) or epigenesis (fig. 3.1).

Maternal inheritance describes the differentiation of the primordial germ cells at the earliest stage of development. The material needed to differentiate the primordial germ cells (i.e., germ plasm) is inherited directly from the mother's oocyte, so everything is there from the point of fertilization on. To illustrate how this works, let's look at the common fruit fly. When

Maternal Inheritance	Epigenesis
Primordial germ cells differentiate at the earliest stages of embryonic development.	Primordial germ cells differentiate at later stages of embryonic development.
Material needed to differentiate primordial germ cells inherited directly from mother via the oocyte.	Cluster of cells in the epiblast receive signals that induce them to become primordial germ cells.

FIGURE 3.1 | Maternal inheritance and epigenesis modes of primordial germ cell differentiation. Created with BioRender.com.

sperm and egg unite during fruit fly fertilization, the result is a single cell, called a zygote. This zygote goes through a process of division in a special way—the nuclei within the cell divide rapidly, but the cell doesn't actually divide. The result is a single, big cell with multiple nuclei, and this is called a *syncytium*. Syncytia are pretty common within insect embryos but don't occur in vertebrate embryos. As these nuclei divide, germ plasm from the ovum that contributed to the zygote is shuffled off into one end of the syncytium where it will eventually become localized within a small number of cells. Thus, in species that undergo maternal inheritance, the primordial germ cells differentiate at the earliest stages of development.

Epigenesis describes the process of differentiating primordial germ cells later in development. It works somewhat differently than maternal inheritance and is more common in vertebrates like us. It's helpful to look at how primordial germ cells originate in mice (*Mus musculus*). Mice, like all other vertebrates, don't have a syncytium. The single-celled zygote that arises from fertilization divides into two daughter cells. Each of those daughter cells divides, and their daughter cells divide, and so on. Cells keep dividing until many cells are present. At this point, the embryo begins to go through a process of moving the cells around to form different layers (a process called *gastrulation*). Just before the onset of gastrulation, a clump of cells situated in the center of the embryo (called the *epiblast*) receives a series of signals from surrounding cells that push them into becoming primordial germ cells. In mice, the primordial germ cells form around 6.5 days post-fertilization. The process of epigenesis thus causes the primordial germ cells to form slightly later in development, and relies on signaling from surrounding cells, not on material (germ plasm) inherited from the mother.

With these two modes of germ cell differentiation in mind, let's return to thinking about continuity. The argument for the continuity of germline between generations often runs along the following lines: there is "preservation of a continuous germ lineage over successive generations."[7] This means that germline connects generations of organisms going back to the evolutionary origin of germ cells (at least within animals). There is certainly historical precedent for thinking about the continuity of germ cells in this way. Weismann's predecessors in germ cell research, Gustav Jaeger and Moritz Nussbaum, considered this to be the case. And, while Weismann abandoned the idea of germ cells as continuous, he believed that the germ plasm fundamentally connected all generations of organisms throughout evolutionary time. Some recent scientists also claim this to be the case. For example, germline biologist Yukiko Yamashita echoes this idea of germ cells stretching over evolutionary time:

> Each of us, as a multicellular organism, was once two germ cells (one from our mother, the other from our father), each of which was once two germ cells in our grandparents, each of which was once. . . . Throughout this journey of the germ cells, they never died or senesced. And the implication of this, although obvious once stated, is that the very existence of each of us can be tracked back to the gonad of somebody (something) that was not *Homo sapiens* or even a mammal, but something like a choanoflagellate.[8]

As we just saw, the lineage of germ cells is broken in organisms that specify their primordial germ cells via epigenesis — following fertilization and a lot of cell division, primordial germ cells arise from somatic cells. Epigenetic germ cell differentiation is thought to be the ancestral form of germ cell differentiation within metazoans (metazoans are animals).[9] Because germ cells arise from somatic cells during epigenesis, and epigenesis was likely the ancestral mode of primordial germ cell differentiation, it is difficult to support the idea that germ cells form an unbroken chain of cells that exists across generations of organisms all the way back to a common ancestor.

The second way in which the notion of continuity of germ cells can be interpreted is *within an organism*. We can break this down further into two possibilities. First, germ cells constitute a continuous lineage beginning with the first cleavage following fertilization. Second, once the primordial germ cells differentiate during development, they form a continuous cell lineage. In both cases, continuity is predicated on the idea that germ cells are separate and distinct from somatic cells, and that somatic cells cannot become a part of this germ cell lineage, but the second option allows for germ cells to arise later in development from somatic cells. Let's briefly look at both of these options. That germ cells can form a continuous cell lineage, beginning at the point of fertilization and the first cleavage of the egg, could be true for some species that use maternal inheritance to differentiate germ cells. It certainly cannot be true for species that use epigenesis. Again, epigenesis is likely the ancestral mode of primordial germ cell differentiation, and is far more prevalent throughout metazoans than maternal inheritance, and so, there are far more somatic cells giving rise to germ cells during development then there are potentially continuous germlines from the point of fertilization arising via maternal inheritance. Therefore, a claim about continuity like this one—that requires cells to form a continuous lineage throughout the life history of an organism—does not get us far in understanding or explaining the relationship between germ cells or between germ cells and somatic cells.

Finally, let's consider the idea that once the primordial germ cells differentiate during development, they form a continuous cell lineage that cannot be broken by somatic cells. This is currently the most common way of interpreting the idea of continuity in relation to germ cells. It requires us to draw a strict distinction between germ cells and somatic cells, such that somatic cells can never become germ cells once the germ cell lineage is established within an organism. This way of viewing germ cells is fundamental to the Weismann Barrier.

THE WEISMANN BARRIER

We have seen how the Weismann Barrier was established and came to be fundamental to geneticists' thinking about heredity throughout the twentieth century. As concepts about the origins and nature of the germ cells shifted within embryology and then developmental biology, it also gained a hold over thinking in these disciplines. Today many (but certainly not all) scientists consider the Weismann Barrier a fundamental tenet of biology. The empirical veracity of the Weismann Barrier, and the underlying notion of germ cell continuity that it maintains, is assumed, and its normative power is what gives us the definitive statement that somatic cells cannot become germ cells once germline originates during development. Consequently, this normative view of the Weismann Barrier is what dictates that germline can only regenerate from germ cells. It is also an assumption that is foundational to all our genome-editing policies, as we will see. As such, we ought to closely inspect the reasoning and evidence that support it. We can achieve such an inspection by looking at alternatives to the relationship between germ cells and somatic cells that the Weismann Barrier mandates, and the evidence that supports these alternatives. This brings us back to the fundamental question: How does germline regenerate?

There are three possible models for germline regeneration, and each has a different relationship with the Weismann Bar-

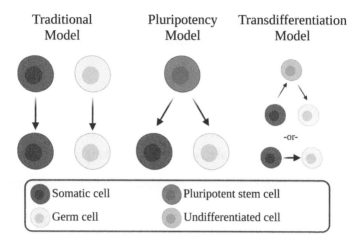

Traditional Model   Pluripotency Model   Transdifferentiation Model

-or-

Somatic cell   Pluripotent stem cell

Germ cell   Undifferentiated cell

FIGURE 3.2 | Three models of germline regeneration. Based on the work of B. Duygu Özpolat. Created with BioRender.com.

rier.[10] Let's explore these three models and the evidence that supports them. The first model is called the *traditional model* (fig. 3.2). It tells us that to regenerate germline, germ cells must be present in the organism. The traditional model, therefore, strictly upholds the Weismann Barrier. It is the model of germline regeneration that most scientists are willing to accept because it adheres to their understanding of and assumptions about the relationship between germ cells and somatic cells. One prominent example of the traditional model is germline stem cell regeneration in fruit flies. When the germline stem cells (the germ cells that sit within a little somatic cell niche within the testes or ovaries) in fruit flies are damaged to the point where they die or leave the stem cell niche, other germ cells that are moving toward differentiation into gametes are supposed to dedifferentiate and fill the stem cell niche, thus regenerating the lost germline stem cells.[11]

The second model is called the *pluripotency model*.[12] Pluripotency refers to the ability of a stem cell to give rise to any other cell type in the body. In this model, you would have a pluripotent stem cell that can give rise to both somatic cells and germ cells. This model has a more complicated relationship with the Weismann Barrier, depending on how we categorize the pluripotent stem cell. If we view stem cells, apart from germline stem cells, as somatic cells, then the pluripotency model of germ cell regeneration rejects the existence of the Weismann Barrier. If, however, we consider pluripotent stem cells to be something distinct from both somatic cells and germ cells, then this model doesn't necessarily contradict the Weismann Barrier, because the Weismann Barrier only addresses germ cells and somatic cells.

The third model is called the *transdifferentiation model*. Transdifferentiation refers to the conversion of a cell from one type to another—for example, from a muscle cell to a liver cell. In this model, you would see a somatic cell become a germ cell. The process could involve either a somatic cell dedifferentiating into a less specialized type of cell (like a stem cell) and then redifferentiating into a germ cell, or it could involve a somatic cell directly transdifferentiating into a germ cell. Either way, the result is the same—a somatic cell becomes a germ cell. This model has a very straightforward relationship with the Weismann Barrier: if it holds, it completely rejects the Weismann Barrier.

What evidence is there to support each of these models? Because I'm interested in challenging the notion of the relationship between germ cells and somatic cells that the Weismann Barrier dictates and the resulting idea that regeneration is limited to the traditional model described above, I'm going to focus on the latter two models of germ cell regeneration: pluripotency and transdifferentiation. To explore the evidence that supports these two models, I have selected a variety of examples meant to highlight the diversity of pluripotency and transdifferentiation across the metazoan clade (fig. 3.3) and showcase some of the best evidence available in support of these models.[13] I should note that germ cell development is not well studied or understood within the vast majority of metazoans, so the evidence remains incomplete and scattered.

PLURIPOTENCY AND GERMLINE REGENERATION

Before we dive into evidence for the pluripotency model of germ cell regeneration, let's lay out some basics about stem cells. Stem cells have two properties: (1) self-renewal, and (2) the ability to differentiate into different cell types (also called *potency*).[14] Self-renewal means that these cells can continuously divide and produce new cells without differentiating. Potency refers to the ability of these stem cells to give rise to cells that can differentiate into different cell types. Scientists categorize stem cell potency in at least four ways: unipotency, multipotency, pluripotency, and totipotency. Unipotency refers to the ability of stem cells to give rise to a very limited number of different cell types—usually just one. Germline stem cells are an example of this kind of potency—they can only give rise to the cell types that will either become sperm or ova. Multipotency refers to the ability of stem cells to give rise

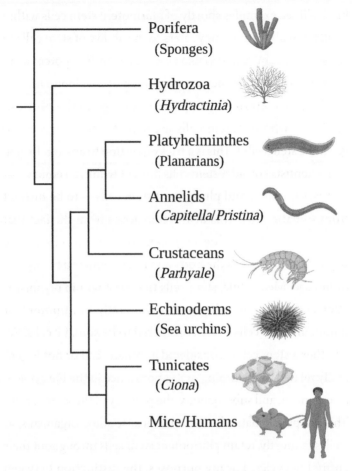

FIGURE 3.3 | Phylogeny of metazoans, indicating the relationships between the different organisms discussed within this chapter. This is not a complete phylogeny of metazoans; it only shows the groups discussed within this chapter. Created with BioRender.com.

to a number of cell types, but these are usually within a closely related family of cells. Scientists consider most adult stem cells to be multipotent. Pluripotency means that stem cells can give rise to nearly all the cell types within the body. Pluripotent stem cells are usually found within the developing embryo, but we'll see examples shortly of pluripotent stem cells within adults. Finally, totipotency refers to the ability of stem cells to construct an entire and viable new organism. Totipotent stem cells are produced at the very beginning of development.

Now, let's return to the question I raised in the previous section—whether stem cells are somatic cells (except for germline stem cells). This is a tricky question to answer. In general, scientists consider stem cells present within an embryo—that is, totipotent and pluripotent stem cells—to be distinct from somatic cells. This distinction arises from the fact that at these early stages of development, these stem cells are the source of all tissues in the body and not committed to any fate or lineage. Meanwhile, stem cells that exist within organisms after embryonic development—generally multipotent or unipotent stem cells—are considered to be somatic cells. So, whether a stem cell is considered a somatic cell or not is generally related to the timing of its appearance in the life cycle of an organism, and subsequently, the potency of those stem cells. This is a problematic distinction, because some organisms, as we'll see shortly, retain pluripotent stem cells throughout their entire life cycles. For my purposes, the distinction between calling stem cells somatic cells or something else only matters insofar as it affects the Weismann Barrier. Thus, although the somatic/non-somatic nature of stem cells is in question, I can say that regardless of where one lands on the question of whether stem cells should be considered somatic cells, the pluripotent model of germ cell regeneration raises problems for the strict distinction and relationship between germ cells and somatic cells that the Weismann Barrier requires.

I am especially interested in pluripotent stem cells—those that can give rise to nearly any cell type within a body. These are present throughout all metazoans during their early development, but they tend to be found only in adults of species that are capable of whole-body regeneration.[15] Whole-body regeneration means that an organism can regrow a completely new body from a small fragment. And, while whole-body regeneration can occur in organisms scattered throughout the metazoan clade, we're going to focus on evidence for the pluripotent model of germ cell regeneration in just a few organisms that are pretty far removed from humans: sponges, *Hydractinia*, and planarians (fig. 3.4). These are marine organisms that live in waters all over the world, and I'll give some details about them as we go.

Sponges are a group of organisms in the phylum Porifera, which sits at the base of the evolutionary tree of metazoans. Sponges are fascinating creatures capable of incredible feats of regeneration; you can put a sponge in a blender and it will regenerate an entire body's structure and functions from the disaggregated cells. Although their bodies lack organ systems (including reproductive organs) and tissues, they are comprised of different cell types, and most of them use sexual reproduction to procreate. For our purposes, two cell types in sponges matter: archeocytes and choanocytes.

Archeocytes are generally agreed to be pluripotent stem cells in sponges—they can differentiate into all the other cell types that make up a sponge. These cells perform all kinds of functions, from aiding in digestion to enabling regeneration. Experiments have shown that without archeocytes, sponges

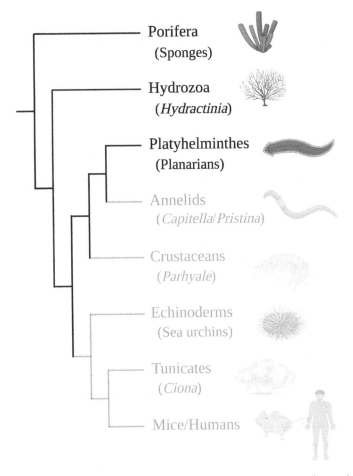

Porifera
(Sponges)

Hydrozoa
(*Hydractinia*)

Platyhelminthes
(Planarians)

Annelids
(*Capitella*/*Pristina*)

Crustaceans
(*Parhyale*)

Echinoderms
(Sea urchins)

Tunicates
(*Ciona*)

Mice/Humans

FIGURE 3.4 | Phylogeny of metazoans, highlighting the positions of Porifera (sponges), Hydrozoa (*Hydractinia*), and Platyhelminthes (planarians). Created with BioRender.com.

are not able to undergo regeneration.[16] Choanocytes are also pluripotent stem cells, and they make up the interior body wall of sponges. They have little tails (called *flagella*) that help pump water throughout the body, while the collars of these cells help trap food and absorb nutrients. Both archeocytes and choanocytes are involved in sexual and asexual reproduc-

FIGURE 3.5 | Sponge (*Aplysina fistularis*, yellow tube sponge). Sponges range in size from under half an inch to over six and a half feet (0.2 centimeters to 2 meters). Photo by Nick Hobgood.

tion in sponges. In sexually reproducing species, the archeo- cytes mostly give rise to oocytes (but sometimes sperm, depending on the species), and choanocytes mostly give rise to sperm (they also give rise to oocytes in some species).[17] During regeneration, archeocytes can give rise to any cell type

FIGURE 3.6 | *Hydractinia* (*Hydractinia symbiopollicaris*) (YPM IZ 096809). Image courtesy of Yale Peabody Museum of Natural History. Created by Eric A. Lazo-Wasem.

in the sponge's body, including choanocytes, and choanocytes can give rise to archeocytes, which can become any cell type.[18] Thus, under both normal and regenerative conditions, gametes originate from pluripotent stem cells (archeocytes and choanocytes) in sponges.

Hydractinia is a genus in the Hydrozoa class that reproduces sexually. You may recall from the discussions of Weismann and Hargitt that hydrozoans are small, predatory marine invertebrates closely related to jellyfish. *Hydractinia* are colonial species found within saltwater environments, and usually live on snail shells. They have tubelike body plans with many types of specialized cells (fig. 3.6). For our purposes, one cell type is important: interstitial cells, also called *i-cells*, which are the stem cells in these organisms. Scientists have been interested in the development and reproduction of hydrozoans for a long time. These were the organisms that caught Weismann's attention in the early 1880s and led him to his understanding of heredity.[19] In Weismann's early studies of hydrozoans, he recognized the i-cells as being somehow involved in the formation of germ cells, and linked these cells loosely with the concept of stem cells (*stammzellen* in German). Weismann did not conceive of stem cells in the same sense that we do today, but he did consider them to be precursors to cell lineages that had the capacity to multiply.[20] Today, scientists consider i-cells in *Hydractinia* to be pluripotent.

In 2004, developmental biologist Werner Müller and colleagues performed a series of experiments that demonstrated the pluripotent abilities of i-cells in *Hydractinia*.[21] They removed the i-cells from a group of *Hydractinia* and then introduced i-cells from donors into these i-cell deficient recipients. Over time, the i-cells from the donors completely took over the recipients, replacing all the somatic cells and the germ cells. Thus, *Hydractinia* i-cells are capable of giving rise to all of the somatic cell lineages as well as germ cells.[22]

Planarians are tiny flatworms in the order Platyhelminthes and live all over the world in saltwater, freshwater, and even terrestrial environments. These organisms are powerhouses of regeneration. For well over a century, scientists have known that if you cut planarians into small pieces, they can regrow entirely new bodies from those little chunks. There are many species of planarians: some reproduce asexually, some sexually, and some

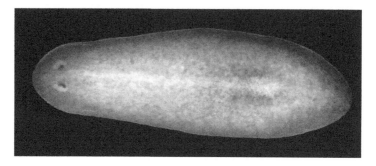

FIGURE 3.7 | Planarian (*Schmidtea mediterranea*). Planarians usually range in length from 0.1 to 0.6 inches (3 to 15 mm), but some species can grow to about a foot long (30 cm). Photo by Alejandro Sánchez Alvarado.

can do both. Planarians have a straightforward body plan—a head and a body (fig. 3.7). The body contains the reproductive organs and the germ cells; the head is completely devoid of germ cells.

In 1901, biologist Thomas Hunt Morgan, who would win a Nobel Prize in 1933 for his work on genetics, cut off the bodies of a bunch of planarians, thereby removing all of the reproductive organs and germ cells, and watched as the remaining heads regrew entirely new bodies, including all the reproductive organs and germs cells.[23] Morgan concluded that the germ cells in these animals could be derived from somatic cells. Scientists later discovered that planarians contain massive populations of cells called *neoblasts*. These neoblasts are pluripotent stem cells that can give rise to both somatic cells and germ cells during regeneration, although recent research indicates that there may be subclasses of neoblasts that have more restricted cell fates.[24]

Taken together, evidence from sponges, *Hydractinia*, and planarians indicates that in some species, especially those that maintain pluripotent stem cells in the adult body, germ cells are readily generated and regenerated from these pluripotent stem cells.

TRANSDIFFERENTIATION AND GERMLINE REGENERATION

Researchers have well documented transdifferentiation, or the ability of cells to change from one type to another.[25] In fact, a lot of cells contain a surprising degree of plasticity in their ability to shift between different cell types. For instance, in *Hydra* (a genus closely related to *Hydractinia*), zymogen cells regularly transdifferentiate into granular mucous cells in the head region.[26] When it comes to transdifferentiation during germ cell regeneration, we need to consider two kinds of evidence. The first is for what I call *natural transdifferentiation*, which means that organisms can transform somatic cells into germ cells *in vivo*, or within their bodies. The second kind of evidence is for what I call *induced transdifferentiation*, which means that somatic cells are capable of transdifferentiation into germ cells *in vitro*—outside the body, typically in the laboratory, in test tubes, or culture dishes.

Natural Transdifferentiation

Let's begin with evidence for natural transdifferentiation by looking at experiments on organisms across the metazoan clade (fig. 3.8), including tunicates (*Ciona intestinalis*), annelids (*Capitella teleta* and *Pristina leidyi*), crustaceans (*Parhyale hawaiensis*), and echinoderms (sea urchins). I'll start with tunicates because they provide the best evidence for natural

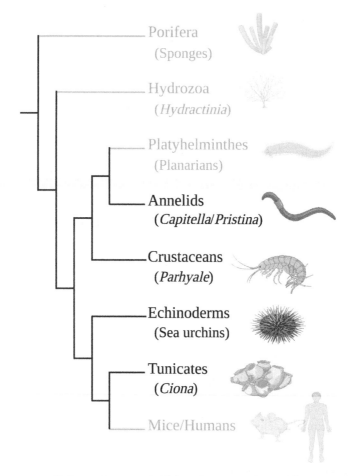

FIGURE 3.8 | Phylogeny of metazoans, highlighting the positions of annelids (*Capitella teleta* and *Pristina leidyi*), crustaceans (*Parhyale hawaiensis*), echinoderms (sea urchins), and tunicates (*Ciona intestinalis*). Created with BioRender.com.

transdifferentiation, and then consider examples in the other groups, because they show that natural transdifferentiation may be fairly widespread among metazoans.

There are around three thousand species of tunicates, marine invertebrates that are filter feeders with a sac-like body

Image by E. A. Lazo-Wasem
Yale Peabody Museum

FIGURE 3.9 | *Ciona intestinalis.* Adult *Ciona* usually grow to be about 6 inches long (15 cm). Image courtesy of Yale Peabody Museum of Natural History. Photo by Eric A. Lazo-Wasem.

structure that is filled with water (fig. 3.9). Their bodies contain two tube-like openings, known as siphons, through which they draw in and expel water. Many tunicates have extensive regenerative abilities, including whole-body regeneration. One species of tunicate that provides excellent evidence for natural transdifferentiation during germline regeneration is *Ciona intestinalis.*

Ciona intestinalis is a solitary tunicate. It basically sits on a substrate in the ocean, like a rock, and filter-feeds from the surrounding environment. *Ciona* are hermaphrodites that release both sperm and eggs into the surrounding water almost simultaneously, but these organisms are self-sterile, meaning that one individual's eggs and sperm cannot fertilize each other.[27] *Ciona*, like most marine organisms, go through a series of different stages before reaching adulthood and their final, tube-like body plan. Juvenile *Ciona* have a basic body plan that includes a head and a tail. In 2002, while developmental biologist Katsumi Takamura and colleagues were investigating the origins of primordial germ cells in *Ciona*, they recognized that the primordial germ cells form in the tail region during the juvenile phase.[28] Further, they discovered that when they removed the tails of *Ciona* juveniles (thus removing the primordial germ cells), the tails would regenerate along with the primordial germ cells and give rise to sexually mature adults.

The question of where the regenerated primordial germ cells came from remained open until 2017, when cell and molecular biologist Keita Yoshida and colleagues performed a series of experiments meant to address this question.[29] Yoshida and colleagues created genetically modified *Ciona* embryos that had different genetic markers for different somatic cell lineages. They removed the tails of these genetically modified juvenile *Ciona* and allowed them to regenerate their tails and develop into reproductively functioning adults. They then genetically sequenced the sperm from these adults, looking for those genetic markers that they had introduced. They found several of those genetic markers for different somatic lineages within the sequenced sperm, including the markers from muscle, neural tissue, and epidermis, indicating that the germ cells had regenerated from these different somatic cell lineages.

The group took their investigation of soma-to-germ trans-differentiation a step further and tested whether *Ciona* would transdifferentiate somatic cells into germ cells even when the primordial germ cells are not removed (i.e., not during germ cell regeneration). They allowed some of their genetically modified embryos to develop to adulthood without removing their tails, and found that among the seven *Ciona* embryos used in this experiment, four had sperm showing somatic expression markers even without removing the primordial germ cells. *Ciona*, therefore, provides strong evidence that (1) transdifferentiation of somatic cells to germ cells is a viable mode of germ cell regeneration, and (2) transdifferentiation of normally somatic lineages to germ cells can happen during normal growth and development. Both of these points call into question how unique the germ cell lineage is within *Ciona*.

Annelids are segmented worms. The annelid clade has habitats that range from terrestrial (e.g., earthworms) to freshwater (e.g., leeches) to marine (e.g., the Pompeii worms that live around hydrothermal vents on the ocean floor) environments. Many annelids have the ability to regenerate anterior segments (those near the head), posterior segments (those near the tail), or both anterior and posterior segments. We're going to look at two species of annelid: *Capitella teleta* and *Pristina leidyi* (fig. 3.10).

 Capitella teleta is a species of marine worm found in sediments along the East and West coasts of the United States. This species is capable of posterior regeneration, including regeneration of its ovaries. In *Capitella*, one cell that appears during early development, called 3D, serves in normal cases as the exclusive progenitor of the germ cells. In 2018, biologists Leah

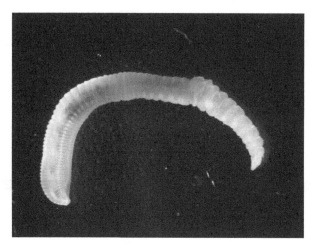

FIGURE 3.10 | *Capitella teleta.* Adult *Capitella* usually range in size from just under half an inch to just over an inch long (1 to 3 cm). Adult *Pristina* are usually less than one-third of an inch long (8 mm). Photo by Lauren Ku.

Dannenberg and Elaine Seaver showed that when the 3D cell is removed, some larvae still develop primordial germ cells.[30] When allowed to continue developing, these larvae turn into adults with fully functional reproductive systems capable of giving rise to offspring. This evidence indicates that within *Capitella*, other cells, which normally give rise to somatic cell lineages, are capable of converting into the germline progenitor cell when it is removed early during development.

Pristina leidyi is a species of freshwater worm found throughout North and South America, and it is capable of regenerating both anterior and posterior segments following amputation. These worms can reproduce both sexually and asexually, and both adult males and females of the species contain gonads where the germ cells are located. The gonads in *Pristina* are located in the anterior segments of the body, and the posterior segments contain a cluster of pluripotent stem cells. In a series of experiments published in 2016, germline biologist B. Duygu Özpolat and colleagues showed that when these animals were starved for a month, their gonads regressed to the point where they could not be found through either their gene expression or by microscopy.[31] Upon refeeding, *Pristina* quickly reestablished their gonads and became reproductively viable. Following up on this experiment, Özpolat and colleagues found that when they amputated the anterior of these worms (removing the gonads) and the posterior (removing segments that contain the pluripotent stem cells), the resulting body segments were capable of regenerating the germ cells and the gonads.[32] From this set of experiments, we can see that in *Pristina*, it is likely that somatic cells can transdifferentiate into germline cells during regeneration.

Crustaceans are invertebrates with a hard exoskeleton and segmented body plan. They range in size from microscopic all the way up to Japanese spider crabs, which can have leg spans of 12.5 feet (3.8 meters). Compared to some of the other species that we've seen thus far, crustaceans tend to have more limited regenerative capabilities, although they are still quite impressive when compared to our own. For instance, Florida stone crabs, which are found throughout the North Atlantic, can regenerate both of their claws. This comes in handy, because harvesting these crabs for food includes removing one or both claws from the live crab and returning the crab to the ocean.

FIGURE 3.11 | *Parhyale hawaiensis.* Adult females can grow to just over three-quarters of an inch long (2 cm). Image courtesy of Museum of Comparative Zoology, Harvard University. Created by Jennifer W. Trimble.

We're going to look specifically at one species of crustacean: *Parhyale hawaiensis* (fig. 3.11).

Parhyale hawaiensis is a species of marine crustacean found in shallow, warm waters worldwide. Much like Florida stone crabs, *Parhyale* are capable of regenerating their limbs. *Parhyale* exhibit early cell fate restriction and are thought to use maternal inheritance to specify their germline. By the eight-cell stage, each cell in the embryo has begun to differentiate. One of these cells, called the *g-cell*, exclusively gives rise to the primordial germ cells. In 2007, biologist Melinda Modrell found that when the g-cell is removed at this early stage of the embryo, the primordial germ cells will not form.[33] Surprisingly these embryos developed into fertile adults that produced normal offspring, indicating that the germline regenerated from somatic cells at some point in the juvenile phase of their life cycle. It is likely that germline regeneration in *Parhyale* is induced by signals

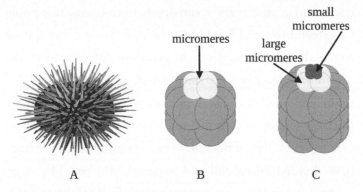

FIGURE 3.12 | (A) Adult sea urchin. (B) 16-cell stage sea urchin embryo. (C) 32-cell stage sea urchin embryo. Note that at the 32-cell stage, the micromeres that appear at the 16-cell stage have divided into small and large micromeres. Created with BioRender.com.

coming from the developing gonads, although this hypothesis has yet to be thoroughly tested.[34]

Echinoderms are marine invertebrates recognizable by the radial symmetry of their body plans. They can be found at any ocean depth and include the starfish, sea urchins, and sand dollars that regularly turn up along beaches. Many echinoderms have strong regenerative capabilities, including regenerating parts (like missing limbs) and even whole-body regeneration. For instance, starfish have the ability to regenerate their arms, and can even regenerate an entirely new body from an arm that retains part of their central disc. We're going to look at one group of echinoderms: sea urchins (fig. 3.12).

Sea urchins are a diverse group of echinoderms characterized by their pentaradial body symmetry (their bodies branch into five distinct compartments) and their spiny outer layer. These spiny creatures can be found on the seabed of every ocean and inhabit every ocean depth, from coastal tide pools to sixteen thousand feet below sea level. Sea urchins appear to use the epigenetic mode of germline differentiation. As the embryo reaches the sixteen-cell stage, it produces a cluster of four small cells called micromeres that sits adjacent to the larger clump of cells that make up the rest of the embryo (see fig. 3.12). By the thirty-two-cell stage, these micromeres have divided into small and large micromeres. The large micromeres rest in contact with the remainder of the embryo, and the small micromeres sit atop these large micromeres like a pearly crown. The small micromeres from the thirty-two-cell stage embryo are the precursors to the primordial germ cells. When the micromeres are removed from sea urchin embryos at the sixteen-cell stage (before the micromeres have divided into large and small), the remaining embryos will nonetheless develop normally and give rise to adults with gametes.[35] It is unclear how sea urchins compensate for the loss of these primordial germ cell-precursor cells, but cells that normally give rise to somatic lineages must be involved.

Induced Transdifferentiation

So far I have focused on cases that provide evidence for natural transdifferentiation. Within this subcategory, we have seen that some organisms readily regenerate their germlines (and in some cases even generate their germlines) from somatic cells. Now let's turn to cases that provide evidence for induced transdifferentiation. In order to understand induced transdifferentiation, we first need a better understanding of the history behind thinking about how cells can change their fates, also called *cell reprogramming.*

Research in the late nineteenth century had given scientists the idea that cells within embryos narrow their fates fairly quickly during development. Some scientists, like the embryologist Wilhelm Roux, thought that cell fates are determined at the earliest stages of the embryo and incapable of change thereafter, whereas others, like the embryologist Hans Driesch, thought that there is more room for cells to change their fates. In the early twentieth century, the understanding emerged that, throughout development, cells gradually lose their potential and progressively arrive at their terminally differentiated fates—as muscle cells, neurons, and so on. Cells that are terminally differentiated have reached their final fate or cell type. Once cells reached these end stages, they were thought to be set in those fates.

In the 1950s, this way of thinking about cells began to change. Two developmental biologists, Robert Briggs and Thomas King, performed a series of experiments in which they removed the nuclei from oocytes and replaced them with nuclei from late-stage embryos and tadpoles.[36] The result was an oocyte capable of some development. While these experiments were ground-breaking, Briggs and King used nuclei from cells that had not quite reached their fully differentiated states. In 1962, developmental biologist John Gurdon replicated these experiments using nuclei from fully differentiated cells harvested from the intestines of tadpoles.[37] The technique that Briggs, King, and Gurdon pioneered, called somatic cell nuclear transfer, showed that differentiated cells retained the potential to adopt other, less differentiated states. Consequently, these were also the first cloning experiments.[38]

In the 1980s, scientists began to find that certain transcription factors, which are proteins produced by the cell that normally help bind it to its specific fate, could also change a cell's fate. In 1987, three scientists, Robert Davis, Harold Weintraub, and Andrew Lassar, showed that fibroblasts (a type of connective tissue cell) could be converted into myoblasts (a type of undifferentiated cell that gives rise to muscle cells), when forced to express a transcription factor called MyoD.[39] Other subsequent researchers found that forced expression of transcription factors could convert primary B and T cells (both are immune system cells) into a different type of immune cell called a *macrophage*.[40] From this research, it became clear that transcription factors could induce differentiated cells to switch their fates. One question remained, though: Can terminally differentiated cells be induced into a stem cell fate?

Recall that stem cells are defined by having two properties: (1) they can self-renew, and (2) they can differentiate into different cell types.[41] Recall also that stem cells can have different levels of ability to differentiate. For instance, pluripotent stem cells, as we saw in sponges, planarians, and *Hydractinia*, can give rise to any cell type in the body.

In 2006, stem cell biologists Kazutoshi Takahashi and Shinya Yamanaka published a series of experiments in which they showed that the introduction of a series of four transcription factors (Oct3/4, Sox2, c-Myc, and Klf4) could induce mouse embryonic fibroblast cells into a pluripotent state *in vitro.*[42] These new cells were called *induced pluripotent stem cells* (iPSCs), and scientists around the world quickly began to fine-tune the system that Takahashi and Yamanaka had developed.[43] For instance, in 2007, a group of researchers showed that iPSCs could be formed from human somatic cells derived from fetal and newborn tissues and from adult tissues.[44] In 2012, the Nobel Prize in Physiology or Medicine was awarded jointly to John Gurdon and Shinya Yamanaka "for the discovery that mature cells can be reprogrammed to become pluripotent."[45] Cell reprogramming then, has a long and storied history within biology.

The history of cell reprogramming that we have just gone through indicates that somatic cells have far more built-in plasticity than scientists often give them credit for. It also shows that somatic cells have the potential to transdifferentiate into other cell types, including pluripotent stem cells. Based on this history, I argue that we should pay close attention to evidence of germline regeneration via induced transdifferentiation, because it can show us that, *given the right conditions,* somatic cells can become germ cells. Further, in the species that I'm going to focus on most, mice and humans, it is incredibly difficult (in the case of mice) or nearly impossible (in the case of humans) to provide evidence for natural transdifferentiation for a number of reasons, including the reproductive biology of these organisms and ethical prohibitions (in the case of humans). So discovering the right conditions for induced transdifferentiation *in vitro* is an important step in determining whether those conditions can happen *in vivo.*

Now let's take a look at the available evidence for induced

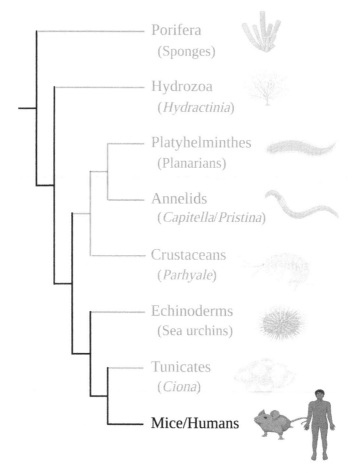

FIGURE 3.13 | Phylogeny of metazoans, highlighting the positions of mice and humans (mammals). Created with BioRender.com.

transdifferentiation. I'm going to focus most on what has been accomplished in mice and in humans (fig. 3.13), but before I do, I need to make an important distinction. Induced transdifferentiation in mice has yielded progeny, whereas all of the work that I describe for humans has not. This distinction is particularly important for two reasons. First, yielding progeny from induced transdifferentiation is considered the gold standard of evidence that the cellular reprogramming of iPSCs (somatic cells) to germ cells works. Second, yielding progeny from the induced transdifferentiation of somatic cells to germ cells in humans has a lot of ethical boundaries that have not, to my knowledge, been crossed.

Mice are one of the most well-studied organisms within biology, and they are thought to be especially useful for understanding biological processes in humans. As discussed previously, mice use epigenesis to differentiate their primordial germ cells just before the time that the embryo enters the process of gastrulation. Unlike many of the organisms I've discussed, mouse gonads are hard to access for *in vivo* studies.

In 2007, just one year after his lab had introduced iPSCs, stem cell biologist Shinya Yamanaka and two other colleagues introduced iPSCs into early-stage mouse embryos, before the time at which primordial germ cells began to differentiate.[46] They allowed these embryos to develop, and the result was chimeric mouse pups—mice that contained the genetic information from the original embryo cells *and* genetic information from the introduced iPSCs. The group then bred some of these chimeric mice and found that the genetic information from the iPSCs appeared in the resulting progeny, indicating that the introduced iPSCs had given rise to germ cells. These experiments served as proof of concept that iPSCs could be integrated throughout a developing embryo and indicated that iPSCs could give rise to germline cells, albeit indirectly.

In 2011, a group of stem cell and developmental biologists led by Katsuhiko Hayashi and Mitinori Saitou showed how iPSCs generated from mouse embryonic fibroblasts could be used to create what they called "primordial germ cell–like cells."[47] The name primordial germ cell–like cells indicates that they operate like primordial germ cells, but they are derived from an artificial technique and thus should be thought of as slightly different from primordial germ cells that form in the body through natural processes. We'll see throughout the next few examples of scientific studies that scientists frequently add "like" to a cell name in order to make this distinction between natural and artificial apparent, and I will follow their naming rules as I explain their science. The group was able to transdifferentiate iPSCs *in vitro* into germ-like cells (i.e., the cells displayed many of the genetic markers of germ cells). They then transplanted these germ-like cells into the testes of infertile male mice. Following the transplant, several of the mice could create sperm. The group harvested some of the resulting sperm and used them to fertilize eggs, which, when transplanted into female mice, gave rise to offspring. This served as a proof of concept that iPSCs could give rise to male germ cells *in vitro*.

Female germ cells (e.g., oocytes) are much harder to recreate than male germ cells (e.g., sperm) because proper development of oocytes relies heavily on the presence and influence of the surrounding (somatic) ovarian tissue. In 2012, another group led by Katsuhiko Hayashi and Mitinori Saitou showed how oocytes could be induced to form *in vitro* from somatic cells.[48] The group harvested somatic cells from mouse embryos, turned them into iPSCs, which they induced into becoming primordial germ cell–like cells. They then combined these primordial germ cell–like cells with somatic cells from mouse ovaries to create what are called "reconstituted ovaries." *Reconstituted* means that the ovaries were reformed or reconstructed from these disparate batches of cells. Within these reconstituted ovaries, the primordial germ cell–like cells matured into oocytes. The group removed these oocytes, fertilized them *in vitro*, transplanted the resulting early-stage embryos into female mice, and found that they gave rise to offspring. In the years since, groups around the world have refined this *in vitro* culturing system and shown that iPSCs derived from adult mice and chickens are also capable of becoming fully functional oocytes.[49]

In mice and in chickens, then, scientists have achieved the gold standard of evidence that induced transdifferentiation of somatic cells into germ cells (via iPSCs) can result in progeny. Now, what about cases where scientists have not yet achieved this gold standard of evidence? In particular, what about humans?

I mentioned previously that mice are often used as models for understanding biological processes in humans. While mice have given us insights into many aspects of human biology and processes related to our germline, there are actually quite a few critical differences between mouse and human germ cell biology that make extrapolating evidence from mice to humans problematic. For instance, just before gastrulation, mouse embryos contain a structure called the *extraembryonic ectoderm* that plays an important role in the formation of primordial germ cells by expressing a protein called BMP4.[50] Humans do not have extraembryonic ectoderm in their pre-gastrulation embryos, nor do human primordial germ cells appear to rely on BMP4 to differentiate. In spite of differences like these, scientists have made substantial strides in turning human iPSCs into germline cells.

In 2015, a large group of scientists, including germline biologist Kotaro Sasaki, as well as Shinya Yamanaka and Mitinori Saitou, derived iPSCs from human blood cells.[51] They cultured these iPSCs *in vitro* and differentiated them into what they called "human primordial germ cell–like cells." However, unlike human primordial germ cells *in vivo*, these cells were missing the expression of several proteins (for example, DAZL) that primordial germ cells only express after they have migrated into the gonads. Thus, this group provided the first evidence that human iPSCs are capable of becoming germ cells, but only up to a fairly early point in germ cell development.

Just three years later, in 2018, another group of biologists, including Chika Yamashiro and, again, Kotaro Sasaki and Mitinori Saitou, pushed human iPSCs into even later stages of germ cell development.[52] The group derived human primordial germ cell–like cells from iPSCs and cultured them *in vitro* with somatic cells harvested from embryonic mouse ovaries. Together, these two types of cells generated what scientists call *xenogenic reconstituted ovaries*, or *xrOvaries*. Like the previously mentioned "reconstituted" ovaries, these were reconstructed by harvesting somatic cells from the ovaries of mice. The twist here is the xenogenic aspect, which means that the ovaries were composed of cells from more than one species—germline cells from humans and somatic cells from mice. After the group let these xrOvaries culture *in vitro* for several months, the human primordial germ cell–like cells gradually acquired the late germ cell markers that had been missing from earlier research. These cells resembled human precursors to oocytes *in vivo* but did not enter meiosis (the type of cell division that produces gametes). Experiments in some other mammals (e.g., pigs and cows) have given similar results—scientists have been able to induce transdifferentiation of somatic cells into germ cells using iPSCs, but have not gotten to the point of producing functional gametes.[53]

Taken together, the experiments showing natural transdifferentiation in tunicates (*Ciona intestinalis*), annelids (*Capitella teleta* and *Pristina leidyi*), crustaceans (*Parhyale hawaiensis*), and echinoderms (sea urchins), and the experiments showing induced transdifferentiation in mice and humans (and chickens, pigs, and cows), provide strong evidence that transdifferentiation from somatic cells to germ cells can occur among all metazoans.

THE FALSE VITALITY OF THE WALKING DEAD? GERMLINE REGENERATION

Let's return to the task that I set out at the beginning of this chapter. Do the assumptions that arise from the history of thinking about germ cells hold as valid, or do they render to our understanding of germ cells the false vitality of the walking dead? Recall that the historical assumptions are that (1) scientists can accurately and reliably distinguish between germ cells and somatic cells; (2) scientists can definitively say that the germ cell lineage is continuous, so that there is an unbroken line of descent among germ cells within an organism;

and (3) scientists can state that under no circumstances can somatic cells become germ cells (the Weismann Barrier).

Let's start with the assumption that scientists can accurately and reliably distinguish germ cells and somatic cells. Individual methods of germ cell identification, like relying solely on gene expression to identify germ cells, can introduce an uncomfortable amount of uncertainty into experiments. The genes used to identify germ cells are not always specific to germ cells, just as the morphological and selective staining criteria in Hargitt's time were not specific to germ cells. Combining methods, like gene expression and cell lineage tracing, can give scientists a much higher degree of certainty that the cells they identify as germ cells are, in fact, germ cells. But doing this is a time- and resource-intensive endeavor, and it is not a common practice in germ cell studies. Given these limitations on scientists' abilities to identify germ cells accurately and reliably, claims that we are certain that germ cells and somatic cells maintain separate and distinct lineages should be treated with caution. Such claims, of course, underlie the other two historical assumptions; thus, the foundation for germ cell continuity and the Weismann Barrier rests on uncertain grounds.

Now, let's look at germ cell continuity. I went through the various ways in which scientists currently conceive of germ cells as continuous. Basing a concept of continuity on germ plasm, as we saw, is highly problematic, as is the idea that there is an unbroken line of descent among germ cells between generations leading back to a common ancestor. There isn't evidence to support that either of these views can be applied throughout metazoans. The remaining concept of continuity—germ cells as a continuous cell lineage within an organism—is also problematic. Recall that I broke this notion down into two further categories: (1) a continuous cell lineage beginning with the first cleavage of the egg following fertilization, and (2) a continuous cell lineage beginning with the differentiation of the primordial germ cells within the embryo. We saw how the first category of continuity here cannot apply to all organisms, because many metazoans form their primordial germ cells well after the first cleavage, using a process called epigenesis (which is also considered to be the ancestral means of differentiating germ cells). The second category is the basis of the Weismann Barrier, and I'll address that shortly. Overall, then, I can say that germ cells form a continuous lineage only insofar as somatic cells in the organism form cell lineages; there doesn't appear to be anything different or special about the germ cell lineage that marks its "continuity" as unique.

Finally, let's look at the Weismann Barrier (and the associated idea of germ cell continuity described above). The Weismann Barrier is often treated as a normative claim, one that defines the way in which germ cells and somatic cells must be related and prohibits somatic cells from becoming germ cells once germline forms during development. Many (but not all) scientists consider this claim to be a universal, impenetrable, and fundamental tenet of biology. And yet, we have seen many examples that contravene (pluripotent stem cells in sponges, *Hydractinia*, and planarians) or break (transdifferentiation in *Ciona*, *Capitella*, *Pristina*, *Parhyale*, sea urchins, mice, and even humans) the Weismann Barrier and the supposed continuity of germ cells once primordial germ cells differentiate during development.

All of this reasoning and evidence lead me to the conclusion that the Weismann Barrier and the notion of an absolutely separate, distinct, and continuous germ cell lineage do, in fact, have the false vitality of the walking dead. These are assumptions without solid enough foundations to make them universally applicable throughout metazoans, and the exceptions to them are so widespread that we should seriously reconsider their usage. This leads me back to my driving question for this book: How does germline regenerate?

Throughout this chapter, we have seen evidence that germline regenerates in a variety of ways—via germ cells, via pluripotent cells, and via transdifferentiation of somatic cells. There are two major outcomes of this more expansive view of germline regeneration and the subsequent relaxation of the normative role of the Weismann Barrier. First, scientists can remove many of the conceptual limitations on the idea of germline regeneration. Now, scientists who have viewed the Weismann Barrier as normative can consider the possibility that stem cells and somatic cells can play a role in germline regeneration and design experiments accordingly. This may sound obvious, but consider: you don't find what you're not looking for. If you consider germ cells to be the only source of germline regeneration, you're probably not going to find evidence for somatic cells or stem cells in this process, nor are you going to design experiments to check whether non–germ cells are becoming part of the germline. Big, important questions arise from this revision in thinking about germline regeneration: for example, How widespread is germline regeneration within metazoans? How has germline regeneration evolved within metazoans? What are the conditions under which germline regeneration occurs?

The second major outcome is that we can revisit other concepts, not related to germline regeneration, that are built on these flawed assumptions and see how reenvisioning the nature of germ cells and their relationship with somatic cells affects them.

4   Implications of Reenvisioning Germline Regeneration

So far we have seen how assumptions about germ cells—how they are identified, how they can be continuous, and the Weismann Barrier—have arisen, become integrated into the life sciences, and underlie a lot of scientific thinking about germline regeneration. We have also seen why we have reason to be skeptical about the validity of these assumptions and the conclusions built on them. A normative view of the Weismann Barrier, which defines the relationship between germ cells and somatic cells in a strict and inviolable way, affects biologists' views of many essential life processes. For instance, consider reproduction. One in eight couples worldwide suffer from infertility, and some of this is due to the inability to make functional gametes. Now think about how species change over time—accreting changes over generations by passing heritable traits down through offspring. Many of our current concepts of evolution and heredity hinge on the Weismann Barrier and its ability to block the inheritance of acquired traits.[1]

Instead of focusing on these important topics, I want to explore the real-world implications of a permeable Weismann

Barrier by looking at human genome editing and show how assumptions about the nature of germ cells have massive repercussions for this issue that sits at the intersection of science and society.

INTRODUCTION TO GENOME EDITING

Humans have been modifying livestock and agriculture to suit their interests for thousands of years. We've bred cows, sheep, goats, chickens, and many other species for particular traits that we value; and we've gone through a similar process with crops like corn, wheat, rye, and rice. While we have undoubtedly modified the genomes of these species by selectively breeding them over such long timespans, this does not count as genome editing in the technical sense, because the effects on the genome have been indirect and not targeted.

More direct attempts to modify genomes began in the early-to-mid-twentieth century, when scientists began to conduct mutation studies to understand how genes are arranged and how they work. These studies also do not quite count as genome editing either; while the intent was to alter genomes, the technologies scientists used, like exposure to radiation, X-rays, and different chemicals, did not allow them to target specific genes or sections of genomes. Genome editing, for my purposes, means the ability to alter directly the genetic material of an organism through the insertion, deletion, repression, or enhancement of genes in a targeted way.

Technologies that have allowed scientists to manipulate the genomes of organisms in targeted ways arose in the 1970s.

These early forms of genome-editing technologies, like recombinant DNA (rDNA), gave scientists direct access to modifying organisms' genomes, but they were time-consuming, inefficient, and expensive, and they required a great deal of specialized knowledge to use.[2] Over the intervening decades, scientists have developed new and more efficient technologies for altering genomes. There are currently three main genome-editing technologies used within the scientific community: zinc finger nucleases (ZFNs), transcription activator-like effector nucleases (TALENs), and clustered regularly interspaced short palindromic repeats (CRISPR). All three of these technologies are based on the same principle of using nucleases and guides to create a double-stranded break in DNA at specific sites, but there are some distinct differences in their cost, efficiency, and ease of use.[3]

ZFNs were first used to alter the genomes of developing frogs in 2001, and after several years, this technology was expanded to be used in mammals.[4] While ZFNs were far faster and more efficient than previous genome-editing technologies, the limits of this technology quickly became apparent. By 2011 scientists had shown that ZFNs were not always good at hitting their desired genome targets.[5] The technology was also initially proprietary, meaning that ZFN kits had to be purchased from specific companies, often at a high cost that put it beyond the scope of many laboratories.

TALENs, meanwhile, were first introduced in 2011, when a group of scientists used this technology to edit rat genomes.[6] While ZFNs are manufactured, TALENs are derived from naturally occurring proteins, so their use and distribution is not proprietary. Although it is cheaper to create TALENs than to purchase ZFN kits, it is a long and cumbersome process to design and generate the desired TALEN for an experiment. The process to verify that the TALEN has hit and modified the correct target is also long and tedious, making this method inefficient to use.

CRISPR was introduced as a tool for molecular biology in 2012 and quickly became the most dominant current genome-editing technology.[7] While CRISPR has its own set of drawbacks, its popularity is grounded in the fact that it is far simpler, cheaper, and faster to use than ZFNs or TALENs. It has become pervasive within the scientific community, and its quick uptake led the journal *Science* to dub CRISPR the "Breakthrough of the Year" in 2015, and the journal *Nature* to call it out as marking the "dawn of the gene-editing age" in 2016.[8] In 2020, the Nobel Prize in Chemistry was awarded to two of the creators of CRISPR, Jennifer Doudna and Emmanuelle Charpentier. The CRISPR genome-editing system sparked a miniature revolution within the scientific community, and some have argued that CRISPR has had a democratizing effect on science, because now even small laboratories can gain enough funding and technical expertise to modify their organisms' genomes for experiments.[9]

While CRISPR has been a boon for modifying genomes in the laboratory, it has also sparked renewed interest in gene therapy—a biomedical intervention to treat disease by modifying specifically targeted genes via disruption, correction, or replacement. The first US application of gene therapy to humans occurred in 1990 to treat ADA SCID (*Adenosine*

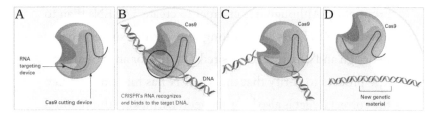

FIGURE 4.1 | CRISPR genome-editing system. (*A*) The CRISPR system has two components joined together: a finely tuned targeting device (a small strand of RNA programmed to look for a specific DNA sequence) and a strong cutting device (an enzyme called Cas9 that can cut through a double strand of DNA). (*B*) Once inside a cell, the CRISPR system locates the DNA it is programmed to find. The CRISPR seeking device recognizes and binds to the target DNA (circled, black). (*C*) The Cas9 enzyme cuts both strands of the DNA. (*D*) Researchers can insert into the cell new sections of DNA. The cell automatically incorporates the new DNA into the gap when it repairs the broken DNA. Images and captions available courtesy of the National Institutes of Health.

Deaminase deficiency Severe Combined Immunodeficiency), an inherited disorder that causes severe immune deficiency and neurological problems.[10] The technology behind early gene therapies required using viruses to deliver therapeutic copies of unmutated genes into patients at random places within their genomes.

Unfortunately, some of the early gene therapy trials had unintended and serious consequences. For instance, in 2000, a group of researchers reported the first successful attempt to edit the malfunctioning gene responsible for a disease called X-SCID (X-linked severe combined immunodeficiency).[11] X-SCID is caused by a mutation in a gene that, under normal circumstances, makes a protein that is necessary for the growth and maturation of the immune system. Children with X-SCID are prone to recurrent infections and, without treatment, do not usually live past infancy. The team successfully introduced a copy of the gene that lacked the mutation responsible for X-SCID into the children in the trial, and the children developed healthy, functioning immune systems. The trial was hailed as a "landmark" for gene therapy.[12] However, just two years after the therapy successfully restored their immune systems, two of the recipients of the treatments developed leukemia apparently caused by the gene therapy. Further research revealed that the therapeutic gene had been integrated into the genome at the location of a gene which, when mutated, causes leukemia. Gene therapies in other, earlier trials had also shown that the use of viruses for gene delivery resulted in severe immune responses in some patients.[13] Cases such as these led to substantial concerns regarding the safety of gene therapies, especially those that rely on viruses to introduce therapeutic genes into random locations in the patients' genome.

The introduction of CRISPR revitalized interest in and applications of gene therapy and changed how they were done. While ZFNs and TALENs were still used for gene therapy, CRISPR quickly supplanted them. CRISPR's low price, ease of use, and precision allow scientists to edit mutated genes *in situ* under controlled conditions rather than introducing therapeutic genes into patients' genomes at random. According to the clinical trials database maintained by the US National Library of Medicine, as of September 2022, there were over 1,300 gene therapy clinical trials recruiting patients or actively underway within the United States. CRISPR has helped gene therapy to explode, and its quick uptake by the biomedical research community has heightened concern for questions of whether, how, and under what circumstances to allow human genome editing.

HUMAN GENOME-EDITING POLICIES AND ETHICS

Debates about human genome editing and gene therapy have been waxing and waning for decades. The ethical debate about allowing alterations of genomes began in the United States in 1981, the same year that scientists published the first germline alteration of mice.[14] That year representatives from three religious groups—Catholic, Jewish, and Protestant—sent a letter to President Jimmy Carter, expressing concern about the growth of genome editing and their fears that these technologies would be used by scientists to alter society.[15] Their concerns, along with those growing within the scientific community led to a governmental report addressing genome editing and gene therapy in 1982, called *Splicing Life.*[16]

This report highlighted the widespread acceptance of the distinction between germ cells and somatic cells, enabling cautious acceptance of somatic cell genome editing. The authors' views on germline genome editing were less clear. *Splicing Life* and other early reports tended to highlight the practical reasons why germline editing would not be needed and was technologically untenable, but they did not draw conclusions about whether, and in what instances, to allow germline genome editing. These debates continued to recur as new genome-editing technologies were introduced. While debates over the ethics and policies of genome editing continued to pop up after 1982, as we have seen, technologies and therapeutic treatments like gene therapy did not become widespread until CRISPR was introduced.

Gene therapies aim to modify particular genes in the body

in targeted ways. Somatic cell modifications, for instance, could be used to introduce modified genes that allow Type II diabetics to process insulin better. Somatic cell gene therapies might also be used in cancer treatments, either to correct identifiable genetic mutations that cause cancers or to introduce genetic modifications that enhance the body's natural abilities to kill wayward cells. In other cases, somatic cell gene therapies aren't enough. It is necessary to edit germline, because the genes have multiple and widespread effects over time, and adding or repressing a single function won't help. Take, for example, Huntington's disease. Huntington's disease is a heritable, progressive, neurodegenerative disorder that affects the cognitive, motor, and psychiatric functions of those affected. We know that Huntington's disease is caused by mutation of a single gene, the *huntingtin* (*HTT*) gene. If we could use CRISPR to alter the mutation in *HTT* that causes Huntington's disease, we could not only slow progression of the disease, but potentially eradicate it. The edits required to cure Huntington's disease would have to be introduced into germline to ensure that the disease would not manifest within an individual or their progeny.[17]

While the early debates about human genome editing were vague with regard to germline genome editing, this topic has come under much more scrutiny recently in response to several events. In the spring of 2015, a group of scientists reported that they had successfully used CRISPR to modify the gene responsible for the fatal blood disorder β-thalassemia in human embryos.[18] These embryos were not viable, but this instance of germline editing served as a proof of concept that gene editing could be accomplished in human embryos and showed obstacles that needed to be overcome before such techniques were used clinically. This report prompted the US National Academy of Sciences and the US National Academy of Medicine to convene an international summit in Washington, DC, on December 1–3, 2015, which was co-hosted by the Royal Society of the United Kingdom and the Chinese Academy of Sciences.[19] Following this summit, the organizing committee released a statement indicating that genome editing of somatic cells could be "appropriately and rigorously evaluated within existing and evolving regulatory frameworks for gene therapy, and [that] regulators [. . . could] weigh risks and potential benefits in approving clinical trials and therapies."[20] The committee further noted that germline genome editing would be "irresponsible" until safety and efficacy issues had been resolved and a broad societal consensus about the appropriateness of the procedure had been achieved.[21] Somatic cell editing thus received the green light from the committee, while germline genome editing was marked as untenable. Concerns about germline genome editing got even more heated in late 2018.

On November 25, 2018, biophysics researcher, Jiankui He, announced the creation of the world's first genetically altered babies.[22] He's announcement came just two days before the start of the Second International Summit on Human Genome Editing, which was another joint meeting of US National Academy of Sciences and US National Academy of Medicine, the Royal Society of the United Kingdom, and the Academy of Sciences of Hong Kong, and was held from November 27–

29 at the University of Hong Kong. Shortly after the second summit, the Organizing Committee released a statement condemning He's work—a sentiment echoed throughout the scientific community.[23] In March of 2019, the director of the World Health Organization stated at an advisory committee meeting on governance and oversight of human genome editing that "it would be irresponsible at this time for anyone to proceed with clinical applications of human germline genome editing."[24] Thus, He's creation of genetically modified babies spurred a proliferation of ethical and policy reports around the world, which now focused heavily on heritable human genome editing.

My aim here is not to give an exhaustive overview of the debates about the ethics and policies surrounding human genome editing. These debates are ongoing, and experts on ethics and science policy have made well-reasoned arguments on all sides.[25] Rather, I want to highlight two important points about all the policies and ethical debates surrounding human genome editing thus far. First, beginning with the release of *Splicing Life*, all the human genome-editing policies and ethical debates have maintained a strict distinction between heritable or germline genome editing and non-heritable or somatic cell genome editing. In other words, they have divided human genome editing into two mutually exclusive categories: somatic cell genome editing and germ cell genome editing. For example, in the National Academies report *Human Genome Editing: Science, Ethics, and Governance*, released in 2017, the authors tell us that "somatic cells contribute to the various tissues of the body but not to the germline, meaning that, in contrast with heritable germline editing, the effects of changes made to somatic cells are limited to the treated individual and would not be inherited by future generations."[26]

Second, in recent years, concerns about somatic cell genome editing have largely fallen into the background as ethicists, scientists, and policymakers have been overwhelmingly concerned with heritable genome editing since late 2018, when He introduced the world's first genetically altered babies. This is evident in the proliferation of genome-editing policies released since 2019.[27] Thus, while germline genome editing is highly controversial, somatic genome editing is much less so, and an enormous amount of research activity is directed at how to target and modify specific genes that cause disease within human non-reproductive cells for biomedical therapies.[28] In fact, recent studies have indicated broad societal support for somatic genome modification to help relieve or reverse genetic diseases like Duchenne muscular dystrophy and cystic fibrosis.[29]

EXPLORING THE REAL-WORLD IMPLICATIONS OF HISTORICAL ASSUMPTIONS

To frame this issue another way, human genome-editing debates and policies have assumed the validity of the Weismann Barrier. They have followed the idea that germ cells and somatic cells are separate and distinct, and that the relationship between the two is inviolable, so policy considerations and ethical debates treat them in separate and distinct ways. Whereas somatic cell genome editing is now lauded for its therapeutic potential, germ cell genome editing is treated much more skeptically and is much more contentious.

This makes sense if researchers trust the validity of the Weismann Barrier, because it maintains that somatic cells are incapable of becoming germ cells. Therefore genome editing applied to somatic cells is not supposed to be heritable. But, as we have seen, there is good reason to doubt the inviolability of the Weismann Barrier, and in fact, it can be quite permeable. A permeable Weismann Barrier means that, *given the right conditions*, somatic cells can, in fact, become germ cells. Now, what does this mean for thinking about human genome editing?

If somatic cells can become germ cells, then the distinction that our human genome-editing policies and ethical debates have held between germ cells and somatic cells collapses. The effect of this collapse should be apparent; if somatic cells can become germ cells, and we are editing somatic cells for medical therapeutics, *we have introduced the potential for inadvertent heritable genome editing.* This may seem like an ethical disaster, or it may seem like a desirable outcome, especially when it comes to devastating heritable diseases. But consider the following hypothetical scenario.

One of the targets of somatic cell genome editing for medical therapeutics is editing genes involved in cancers. One such gene, *PD-1*, produces a protein that helps control the immune system by modulating the activities of T-cells (immune system cells). Certain cancers take advantage of *PD-1*'s role in the immune system. These cancer cells use proteins called *PD-L1* to bind to *PD-1* positive cells and suppress their proliferation, while also inducing cell death in certain T-cells.[30]

In October 2020, three registered clinical trials in the United States were testing the efficacy of treating different cancers by using genome editing to knock out the *PD-1* gene in somatic cells. When it works, this therapy allows the body to produce T-cells that the cancer cells cannot suppress, and can lead to these immune cells killing the cancer cells. While *PD-1* is implicated in some cancers, it also plays a major role in normal immune system responses and autoimmunity. If cells with the *PD-1* gene knocked out were to become germ cells, any children conceived from this genome-edited cell would likely have severe autoimmune issues. *PD-1* deficient mouse models give us an indication of just how problematic this could be. Mice deficient in *PD-1* develop, among other things, lupus-like symptoms, cardiomyopathy (a disease of the heart muscle that makes it harder for the heart to pump blood to the rest of the body), and extraordinary sensitivity to tuberculosis. They have dramatically-reduced survival rates compared to mice that are not deficient in *PD-1*.[31]

The probability of conversions of somatic cells to germ cells is likely to be very low in humans. However, our current human genome-editing policies and ethical debates are not equipped to consider even the *prospect* of inadvertent heritable genome editing. They are currently too rigidly tied to the ideas that germ cells and somatic cells are separate and distinct and that the relationship between these kinds of cells aligns with that prescribed by the Weismann Barrier.

Given the well-founded and widespread reluctance to introduce heritable genomic alterations, this set of assumptions is highly problematic, because our degree of uncertainty and risk is unknown. We do not know how common soma-to-germ transitions are among metazoans, let alone humans. We do not know the conditions under which somatic cells can become germ cells. In large part, we do not know these things because of the normative view that many scientists have held and continue to hold about the Weismann Barrier. In other words, unchallenged thinking about germline as inviolable has led many scientists not to consider the alternative—that somatic cells can become germ cells. Because the risk of soma-to-germ transitions is unknown, our ability to make evidence-based judgements and informed-consent decisions about the effects of somatic cell genome-editing therapies is impossible. So, where do we go with human genome editing from here?

The potential therapeutic benefits of human genome editing are too great to ignore or to put on hold. The evidence for soma-to-germ transdifferentiation that I have presented from *in vivo* and *in vitro* experiments should not interfere with the *exploration* of the therapeutic potential of somatic cell genome editing. Uncertainty and risk are, after all, a component of all biomedical (and scientific) endeavors. But we need to take seriously the risks of soma-to-germ transdifferentiation— something that has not yet been thoroughly investigated within humans or proposed as an issue to be considered in establishing human genome-editing policies. Taking these risks seriously means investing effort and resources into understanding the probabilities of such cellular transitions and, especially, understanding the conditions under which they can occur. Only with this kind of information in hand can scientists and policymakers determine whether our current human genome-editing policies are sufficient to accommodate the risks exposed by reenvisioning the relationship between germ cells and somatic cells. And if our policies are not sufficient to accommodate the risks, we need to revise them.

We could view the permeability of the Weismann Barrier as a scary oversight in our vast and sprawling debates and policies about human genome editing. While this is true, I argue that we would be best served to think of it as a place for further exploration. Rather than halting a multi-billion dollar (annually within the United States) quest for therapeutics that range from treating heritable diseases to cancers, we can treat this realization as a way to open new avenues for better understanding our bodies, the cells that make them up, and the processes that allow us to create future generations. Instead of starting from the idea that somatic cells can never become germ cells, we can start from a new question: How, and under what conditions, can somatic cells become germ cells in humans? In other words, How does germline regenerate?

Epilogue

When you want to remove a particularly tenacious plant, you have to dig it up at the roots. When you want to challenge a widespread assumption about how the world works, you should do the same; expose the historical reasoning and evidence that keep the assumption rooted in scientific thinking. History and philosophy are critical for evaluating current science, and deploying them in combination with scientific analysis is a powerful way to transform our understanding of the natural world.

Using history and philosophy to expose the flaws in assumptions that underlie something as important as thinking about germline regeneration can have far-reaching and unexpected effects. Human genome editing is built on a foundation of conceiving of somatic cells and germ cells as belonging to strictly separate categories. Policies and ethical frameworks surround each of these categories and treat them separately. Throughout this book, however, I have shown that these categories are not as neat and separate as researchers and policymakers believe. Assuming that these categories are mutually exclusive has led us to human genome-editing policies that introduce the possibility of inadvertent heritable genome editing and yet provide no means of preventing it—something that should concern us all. To be absolutely clear, I am not saying that inadvertent heritable genome editing is a likely outcome of somatic cell genome editing. I am saying that it is a possible outcome, and that we do not know the probability of its being an actual outcome. Nor will we know unless and until we challenge our assumptions about the relationships between somatic cells and germ cells and germline regeneration.

Using the history and philosophy of science to challenge assumptions is not limited to the issue of human genome editing. It is a powerful toolkit that can be deployed throughout science, and in other areas of inquiry, and its usage can help us reshape the ways in which we view the world. This book, then, is about germline regeneration, but it is also about how science works, how history shapes current science, how science can shape the practice of history, and how things that we take for granted can have far-reaching effects.

Acknowledgments

This project would not have been possible had not Susan Fitzpatrick, former president of the James S. McDonnell Foundation, seen the value in bringing scholars in history and philosophy of science together with practicing biologists. She was the driving force behind establishing the McDonnell Initiative at the Marine Biological Laboratory in Woods Hole, Massachusetts, which allowed for such an unorthodox book to come into being.

The core working group leaders of the McDonnell Initiative, including B. Duygu Özpolat, Kathryn Maxson Jones, Jennifer Morgan, Lucie Laplane, Michel Vervoort, Eve Gazave, Andrew Inkpen, Ford Doolittle, Frederick Davis, James Collins, and especially Jane Maienschein, were all instrumental in developing the ideas contained in this book. They are all brilliant scholars and, more importantly, brilliant people.

Scholars are like slime molds: our ideas come to fruition when we come together. Thus, such works are the result of communities, not individuals. I would therefore like to thank attendees at various presentations of the materials contained herein, including audiences at the History of Science Society, the Philosophy of Science Association, the Philosophy in Biology and Medicine Initiative, the International Society for History, Philosophy, and Social Studies of Biology, the History of Biology Seminar at the Marine Biological Laboratory, and the Indiana University History and Philosophy of Science Colloquium. Additionally, I would like to thank Paige Madison for her thoughtful feedback on a late draft of this manuscript, and the wonderful staff members of the Marine Biological Laboratory and Arizona State University without whom my work would have been impossible. This especially includes Jessica Ranney, who once told me, "Tell me something smart to say and then you can quote me." Her requirement to appear within all acknowledgments sections is herein met. And, finally, my wonderful editor at the University of Chicago Press, Joseph Calamia, whose patience and helpful feedback ensured that this book came to term.

In addition to professional communities, my personal community made this book possible. I would therefore like to dedicate this book to three incredible women.

First, this book is dedicated to my friend and mentor Jane Maienschein. Her indomitable optimism made me never give up, her willingness to spend long hours reading and discussing drafts of this book meant that I always had the best sounding board in academia, and her constant refrain of "so what?" forced me to be clear about why thinking about germ cells is so important.

Second, this book is dedicated to my friend and collaborator Duygu Özpolat. She spent countless hours teaching me about germ cells; she developed the framework for germ cell regeneration in chapter 3, and she worked with me to think through what continuity means in the context of germ cells. She is a uniquely brilliant, generous, and kind human being whose combination of scientific mind and artistic skills is unparalleled.

Finally, and most importantly, this book is dedicated to my partner in crime, my wife, and the most amazing person I have ever known, Challie Facemire. Her perseverance in the face of adversity continually inspires me to keep going even when I don't know how. She is also the reason I'm still sane enough to write a book even when my mind is full of germ cells.

Notes

INTRODUCTION

1. Maienschein and MacCord, *What Is Regeneration?*
2. Johnson et al., "Germline stem cells"; Johnson et al., "Oocyte generation"; White et al., "Oocyte formation."

CHAPTER 1

1. Leeuwenhoek, "Observationes D. Anthonii Lewenhoeck."
2. In the seventeenth, eighteenth, and nineteenth centuries, the most common terms for female gametes were *ova* or *ovi* (plural) and *ovum* (singular), whereas the most common terms for the male gametes were *spermatozoa* (plural) and *spermatozoon* (singular). Throughout this history, I'll attempt to use the language of science of the time.
3. Note that Baer worked at the University of Tartu which was, at the time, part of the Russian Empire. This area is now part of modern-day Estonia. Throughout the text, I refer to locations according to the modern geopolitical boundaries in which investigators worked at the time under discussion. For more detailed information on the discovery of ova, see Sarton, "Discovery of the mammalian egg"; Baer and O'Malley, "Genesis of the ovum."
4. Waldeyer, *Eierstock und Ei*; Van Beneden, "Composition et la signification de l'œuf"; Churchill, "Weismann's continuity of the germ-plasm."
5. Waldeyer worked at the University of Breslau, which was, at the time, a part

of Prussian Silesia (and in 1871 became a part of the German Empire). This area is now a part of modern-day Poland.

6. For further details about the early discussion of the origin of reproductive (or germ) cells, see Churchill, "Weismann's continuity of the germ-plasm."

7. Darwin, *Origin of Species.*

8. For more on the reception of Darwin's ideas about evolution, see Bowler, *Non-Darwinian Revolution.*

9. Lamarck, *Philosophie zoologique.*

10. Darwin, *Variation of Animals and Plants.*

11. Many historians and philosophers have written about Darwin's theory of pangenesis. For more information on pangenesis and its reception, see Browne, *Charles Darwin*; Endersby, "Darwin on generation"; Geison, "Darwin and heredity"; Olby, "Charles Darwin's manuscript"; Stanford, "Darwin's Pangenesis"; Winther, "Darwin on variation and heredity."

12. Holterhoff, "History and reception."

13. Brooks, *Law of Heredity*; De Vries, *Intracellular Pangenesis.*

14. Churchill, *August Weismann.*

15. Weismann's lecture became chapter 8, titled "The Supposed Transmission of Mutilations," in Weismann, *Essays upon Heredity.*

16. Weismann, *Die Entstehung der Sexualzellen.* For more details about Weismann's research and findings, see Churchill, "Weismann, Hydromedusae."

17. Weismann, *Germ Plasm.*

18. For a brief review of the history of cytology and chromosomes preceding Weismann's germ plasm theory, see Churchill, "August Weismann."

19. In fact, Jaeger also believed in a continuity of germ plasm, but germ plasm could only be found within germ cells. For our purposes, we can lump him together with Nussbaum. For more information about Nussbaum and Jaeger and their thoughts on germ cell continuity, see Churchill, "Weismann's continuity of the germ plasm"; Yamashita, "On the germ-soma distinction"; Robinson, *Prelude to Genetics.*

20. Churchill, "Weismann, Hydromedusae."

21. Wilson, *Cell in Development and Inheritance*, 306–7.

22. Churchill, "August Weismann"; Churchill, *August Weismann.*

23. Wilson, *Cell in Development and Inheritance.*

24. For more on Wilson's importance in the history of cell/experimental biology, see Maienschein, "Shifting assumptions"; Maienschein, *Transforming Traditions.*

25. While Weismann's *The Germ Plasm* was translated to English the year following its initial publication in German, Wilson's *The Cell in Development and Inheritance* was not translated into German, despite going through

three editions. This is particularly interesting because German was considered a common scientific language and understood by many investigators in the United States and the United Kingdom. The phrase "scientific community" thus refers here to an international readership of scientists able to read English.

26. Wilson, *Cell in Development and Inheritance*, 11.

27. For a more extensive look at the conceptual evolution of diagram's purporting to show Weismann's theory, see Griesemer and Wimsatt, "Picturing Weismannism." For a more complete review of Wilson's diagrams, see Maienschein, "From presentation to representation."

28. Whitman, *Embryology of Clepsine*. For a more in-depth view of Whitman's cell lineage study, see Maienschein, "Cell lineage."

29. See Maienschein, "Cell lineage"; Maienschein, *Transforming Traditions*. Cell lineage traditions focused on similar problems also arose in other, European contexts, most notably in Germany, following the works of Theodor Boveri and Oskar Hertwig in the early 1890s. Because our understanding of the Weismann Barrier arose out of the context of the American scholarly community, I focus on Whitman's cell lineage tradition for the sake of accessibility. For further information about early cell lineage studies in the German tradition, see Dröscher, "Images of cell trees."

30. For a more detailed understanding of how embryology changed during this period, see Maienschein, *Transforming Traditions*.

31. Berill and Liu, "Germplasm, Weismann, and hydrozoa."

32. Mendel, "Versuche über pflanzen-hybriden."

33. Many historians have traced different aspects of the origins of the field of genetics; the following texts are offered as merely the tip of the iceberg. Olby, "Emergence of genetics"; Bowler, *Mendelian Revolution*.

34. Much has been written about the history of genetics during the early twentieth century. The following texts are just a small selection of a much larger corpus: Olby, "Dimensions of scientific controversy"; Provine, *Theoretical Population Genetics*; Richmond, "Women in the early history of genetics."

35. Morgan et al., *Mechanism of Mendelian Heredity*. Morgan was not the first to embrace chromosomes as the bearers of the units of heredity, but his research program on *Drosophila* genetics quickly became a dominant source of ideas, investigators, and materials for studying the role of the chromosomes in heredity. For more on Morgan's lab, see Allen, *Thomas Hunt Morgan*; Kohler, *Lords of the Fly*.

36. Simpson, Pittendrigh, and Tiffany, *Introduction to Biology*, 281 (emphasis in original).

37. For more information on the Modern Synthesis, see Burian, "Challenges to

the evolutionary synthesis"; Borrello, "Synthesis and selection"; Dietrich, "Richard Goldschmidt's 'Heresies'"; Mayr and Provine, eds., *Evolutionary Synthesis*; Smocovitis, "Unifying biology"; Sober, "Modern Synthesis."

38. Mayr, *Growth of Biological Thought*, 700.

39. For more about how interpretations and misinterpretations of Weismann's work have been viewed as foundational to the field of genetics, see Mayr, *Growth of Biological Thought*; Mayr and Provine, *Evolutionary Synthesis*; Churchill, "August Weismannn"; Buss, *Evolution of Individuality*; Robinson, *Prelude to Genetics*; Robinson, "August Weismann's Hereditary Theory."

CHAPTER 2

1. For an excellent review of thinking about the origins of germ cells in the early twentieth century, see Heys, "Problem of the origin of germ cells."

2. Heys, "Problem of the origin of germ cells."

3. See the separate bibliography entries for Hargitt, "Germ cells of Coelenterates. I, II, III, V, and VI."

4. Hargitt, "Germ cells of Coelenterates. VI," 15.

5. Heys, "Problem of the origin of germ cells."

6. Hargitt, "Germ cell origins in the adult salamander."

7. Hargitt, "Formation of the sex glands. 1"; Hargitt, "Formation of the sex glands. 2."

8. For details on Hargitt's samples and methods, see Hargitt, "Formation of the sex glands. 1."

9. Hargitt, "Formation of the sex glands. 1," 531.

10. For extensive details on these criteria, see: Hargitt, "Formation of the sex glands. 1."

11. Rubaschkin, "Über die Urgeschlechtszellen."

12. For extensive details on these criteria, see Hargitt, "Formation of the sex glands. 1," 531.

13. Rubaschkin, "Zur Lehre von der Keimbahn."

14. Simkins, "On the Origin and Migration."

15. Hargitt, "Formation of the sex glands. I," 530.

16. Hegner, "Removing the germ-cell determinants."

17. Hegner, "Origin and early history"; Hegner, "Germ-cell determinants"; Hegner, "Eggs of Chrysomelid beetles."

18. Hegner, "Germ-cell determinants," 392 (italics in original).

19. Weismann, "Die Entwicklung der Dipteren."

20. Metschnikoff, "Ueber die Entwicklung" (note that Metschnikoff is spelled "Mecznikoff" in the journal); Leuckart, "Die ungeschlechtliche Fort-pflanzung." For more details, see Yamashita, "On the germ-soma distinction."

21. Hegner, "Removing the germ-cell determinants," 21.

22. Hegner, *Germ-Cell Cycle in Animals*; Hegner, "Studies on germ cells." For more information about the switch, see Yamashita, "On the germ-soma distinction."

23. Hegner, *Germ-Cell Cycle in Animals*, 298

24. Hegner, *Germ-Cell Cycle in Animals*, 296.

25. Hegner, *Germ-Cell Cycle in Animals*, 292.

26. Bounoure, "Recherches sur la lignée germinale."

27. Bounoure, *L'Origin des Cellules Reproductrices*. This was the first book in a two-part series. The second is Bounoure, *Continuité germinale*. Unfortu-nately, no English translation has ever been produced for either of these texts.

28. See Huxley, "Some war-time biological books"; Gurchot, "L'origine des cellules reproductrices."

29. Hargitt, "What is germ plasm?," 345.

30. Hegner, *Germ-Cell Cycle in Animals*, 296.

31. Huxley, "Some war-time biological books," 612.

32. Berrill and Liu, "Germplasm, Weismann, and Hydrozoa," 130.

33. Berrill and Liu. "Germplasm, Weismann, and Hydrozoa," 124

34. Eddy, "Germ plasm."

35. Eddy, "Germ plasm," 230

36. Eddy, "Germ plasm," 230

37. Eddy, "Germ plasm," 263

CHAPTER 3

1. I have excluded the nationalities and/or locations of the scientists dis-cussed in this chapter. I have done this because current scientific fields and even research programs are more likely to involve larger numbers of scientists, scientists from a diverse set of nationalities, and scientists across institutions within or across nations than the historical research discussed in previous chapters. Such diversity is wonderful, but it makes it tricky to accurately identify individuals' nationalities.

2. Scientists have practiced odd and often discriminatory naming con-ventions for genes. For instance, *Piwi* is an abbreviation of "P-element

induced *wimpy* testis in *Drosophila*," and is named for the observed severe sperm-forming defects in male *Drosophila*. *Tudor* mutations were found to be lethal for offspring, and the name is a reference to the Tudor King Henry VII and the several miscarriages experienced by his wives.

3. Note that this discussion of continuity is based on the work of Özpolat and MacCord, and that B. Duygu Özpolat played a critical role in thinking through how continuity of germline can be conceived. See MacCord, and Özpolat, "Is the germline immortal?"

4. Kerr and Dixon, "Ultrastructural study of germ plasm."

5. See, for instance, Gao and Arkov, "Next generation organelles"; Leclère et al., "Maternally localized germ plasm."

6. *Oxford English Dictionary* online, s.v. "germline," accessed September 1, 2022.

7. Gartner, Boag, and Blackwell, "Germline survival and apoptosis."

8. Yamashita, "Unsolved problems in cell biology."

9. Extavour and Akam, "Germ cell specification."

10. The three models of germ cell regeneration were developed by B. Duygu Özpolat.

11. Brawley and Matunis, "Regeneration of male germline"; Kai and Spradling, "Differentiating germ cells."

12. For an interesting take on the relationship between stem cells and germline, see Solana, "Closing the circle."

13. Special thanks are due here to B. Duygu Özpolat for introducing me to many of the examples of germline regeneration provided within this chapter. The evidence in this chapter is focused strictly on metazoans, leaving out the question of how plant germline may be considered. For more information on the question of the uniqueness of plant germlines, see Lanfear, "Do plants have a segregated germline?"

14. Laplane, *Cancer Stem Cells*.

15. Personal communication with Eve Gazave, Lucie Laplane, and Michel Vervoort.

16. Funayama, "Stem cell system."

17. Funayama, "Stem cell system."

18. Funayama, "Stem cell system."

19. Weismann, *Die Entstehung der Sexualzellen*; Weismann, *Germ Plasm*.

20. Frank, Plickert, and Müller, "Cnidarian interstitial cells."

21. Müller, Teo, and Frank, "Totipotent migratory stem cells."

22. Gahan, Bradshaw, Flici, and Frank, "Interstitial stem cells."

23. Morgan, "Growth and regeneration."

24. Newmark, Wang, and Chong, "Germ cell specification"; van Wolfswinkel, Wagner, Reddien, "Single-cell analysis."

25. Gurdon and Byrne, "Nuclear transplantation."

26. For instance, see Siebert, Anton-Erxleben, and Bosch, "Cell type complexity."

27. Sawada, Morita, and Iwano, "Self/non-self recognition mechanisms."

28. Takamura, Fujimura, and Yamaguchi, "Primordial germ cells."

29. Yoshida et al., "Enhanced mutagenesis."

30. Dannenberg and Seaver, "Regeneration of the germline."

31. Özpolat, Sloane, Zattara, and Bely, "Plasticity and regeneration of gonads."

32. Özpolat et al., "Plasticity and regeneration."

33. Modrell, "Early cells fate specification."

34. Kaczmarczyk, "Germline maintenance."

35. Yajima and Wessel, "Small micromeres"; Voronina et al., "Vasa protein expression."

36. Briggs and King, "Transplantation of living nuclei"; King and Briggs, "Transplantation of living nuclei."

37. Gurdon, "Developmental capacity of nuclei."

38. For more on the history of cloning experiments, see Crowe, *Forgotten Clones*; Gurdon and Byrne, "History of cloning"; Maienschein, "On cloning"; Maienschein, "Regenerative medicine's historical roots"; Maienschein, *Whose View of Life?*

39. Davis, Weintraub, and Lassar, "Expression of a single transfected cDNA."

40. Stadtfeld and Hochedlinger, "Induced pluripotency."

41. Laplane, *Cancer Stem Cells.*

42. Takahashi and Yamanaka, "Induction of pluripotent stem cells."

43. Takahashi et al., "Induction of pluripotent stem cells"; Stadtfeld and Hochedlinger, "Induced pluripotency."

44. Yu et al., "Induced pluripotent stem cell lines"; Takahashi et al., "Induction of pluripotent stem cells."

45. Press release. NobelPrize.org. Nobel Prize Outreach AB 2022. 3 June 2022; online at https://www.nobelprize.org/prizes/medicine/2012/press -release/.

46. Okita, Ichisaka, and Yamanaka, "Germline-competent induced pluripotent stem cells."

47. Hayashi et al., "Reconstitution of the mouse germ cell."

48. Hayashi et al., "Offspring from oocytes."

49. Hikabe et al., "Reconstitution in vitro"; Zhao et al., "Production of viable chicken."

50. Makar and Sasaki, "Roadmap of germline development."
51. Sasaki et al., "Robust in vitro induction." These findings were confirmed the same year; see Irie et al., "SOX17 is a critical specifier."
52. Yamashiro et al., "Generation of human oogonia."
53. do Nascimento Costa et al., "Expression of markers"; Dyce, Wen, and Li, "In vitro germline potential"; Dyce et al., "Analysis of oocyte-like cells."

CHAPTER 4

1. See, for instance, Godfrey-Smith's use of the germ/soma distinction to define Darwinian populations in his *Darwinian Populations*.
2. For more on the history, usage, and impact of rDNA, see Wright, "Recombinant DNA technology"; Rabino, "Impact of activist pressures"; Cook-Deegan, *Gene Wars*; Hughes, "Making dollars out of DNA"; Yi, "Who owns what?"
3. Adli, "CRISPR tool kit."
4. Bibikova et al., "Stimulation of homologous recombination."
5. Mussolino and Cathomen, "On target?"
6. Tesson et al., "Knockout rats."
7. Jinek et al., "Programmable dual-RNA–guided DNA."
8. Travis, "Breakthrough of the Year"; See also "CRISPR Everywhere" (editorial).
9. Ledford, "CRISPR, the disruptor."
10. For more on the history of gene therapy, see Wirth, Parker, and Ylä-Herttuala, "History of gene therapy."
11. Cavazzana-Calvo et al., "Gene therapy."
12. Kohn, "Gene therapy for XSCID."
13. Raper et al., "Fatal systemic inflammatory response."
14. Gordon and Ruddle, "Germ line transmission."
15. Cook-Deegan, "Germ-line gene therapy."
16. President's Commission for the Study of Ethical Problems in Medicine, *Splicing Life*.
17. For more insights about heritable modification beyond the germline, see Lewens, "Blurring the germline."
18. Liang et al., "CRISPR/Cas9-mediated gene editing."
19. For a brief history of events surrounding policy summits on germline genome editing, see Hurlbut, "Limits of responsibility"; Bayliss, *Altered Inheritance*.

20. National Academies of Sciences, Engineering, and Medicine, *International Summit on Human Gene Editing*, 7.

21. National Academies of Sciences, Engineering, and Medicine, *International Summit on Human Gene Editing*, 7.

22. Marchione, "Chinese researcher."

23. National Academies of Sciences, Engineering, and Medicine, "Second international summit on human genome editing"; National Academies of Sciences, Engineering, and Medicine, *Heritable Human Genome Editing*; available from https://www.ncbi.nlm.nih.gov/books/NBK535994/ doi: 10.17226/25343, 2020.

24. World Health Organization, "Statement on governance."

25. See, for instance, Jasanoff, Hurlbut, and Saha, "CRISPR democracy"; Hurlbut, "Limits of responsibility"; Hurlbut, *Experiments in Democracy*; Hurlbut, "Human genome editing"; Angrist et al., "Reactions to the National Academies/Royal Society report"; Lewens, "Blurring the germline."

26. National Academies of Sciences, Engineering, and Medicine, *Human Genome Editing*, 84.

27. National Academies of Sciences, Engineering, and Medicine, "Second International summit on human genome editing"; National Academies of Sciences, Engineering, and Medicine. *Heritable Human Genome Editing*.

28. Perry et al., "Genome editing"; Schwank et al., "Functional repair"; Schumann et al., "Generation of knock-in primary human T cells"; Su et al., "CRISPR-Cas9."

29. Hendriks et al., "Reasons for being in favour"; McCaughey et al., "A global social media survey"; Scheufele et al., "US attitudes"; Wang et al., "Public attitudes."

30. For more on the PD-1/PD-L1 pathway in cancer research, see Han, Liu, and Li, "PD-1/PD-L1 pathway."

31. Nishimura, "Nose M Hiai"; Nishimura et al., "Autoimmune dilated cardiomyopathy"; Okazaki et al., "Autoantibodies against cardiac troponin I"; Wang et al., "Establishment of NOD-Pdcd1-/-mice"; Lázár-Molnár et al., "Programmed death-1."

Bibliography

Adli, Mazhar. "The CRISPR tool kit for genome editing and beyond." *Nature Communications* 9, no. 1 (2018): 1–13.

Allen, Garland. *Thomas Hunt Morgan: The Man and His Science.* Princeton, NJ: Princeton University Press, 1978

Angrist, Misha, Rodolphe Barrangou, Françoise Baylis, Carolyn Brokowski, Gaetan Burgio, Arthur Caplan, Carolyn Riley Chapman, et al. "Reactions to the National Academies/Royal Society report on heritable human genome editing." *CRISPR Journal* 3, no. 5 (2020): 332–49.

Baer, Karl Ernst von, and Charles Donald O'Malley. "On the Genesis of the Ovum of Mammals and of Man." *Isis* 47, no. 2 (1956): 117–53.

Bayliss, Françoise. *Altered Inheritance.* Cambridge, MA: Harvard University Press, 2019.

Berill, Norman John, and Chien-Kang Liu. "Germplasm, Weismann, and Hydrozoa." *Quarterly Review of Biology* 23, no. 2 (1948): 124–32.

Bibikova, Marina, Dana Carroll, David J. Segal, Jonathan K. Traut-

man, Jeff Smith, Yang-Gyun Kim, and Srinivasan Chandrase-
garan. "Stimulation of homologous recombination through
targeted cleavage by chimeric nucleases." *Molecular and Cellular
Biology* 21, no. 1 (2001): 289–97.

Bline, Abigail P., Anne Le Goff, and Patrick Allard. "What is lost in
the Weismann Barrier?" *Journal of Developmental Biology* 8, no.
4 (2020): 35.

Borrello, Mark E. "Synthesis and Selection: Wynne-Edwards'
Challenge to David Lack." *Journal of the History of Biology* 36,
no. 3 (2003): 531–66.

Bounoure, Louis. "Recherches sur la lignée germinale chez la gre-
nouille rousse aux premiers stades du développement." *Annales
des sciences naturelles Zoologies* 17 (1934): 67–248.

———. *L'Origin des Cellules Reproductrices et la Probleme de la
Lignée Germinale.* Paris: Gauthier-Villars, 1939.

Bowler, Peter J. *The Non-Darwinian Revolution: Reinterpreting a
Historical Myth.* Baltimore, MD: Johns Hopkins University
Press, 1988.

———. *The Mendelian Revolution: The Emergence of Hereditarian
Concepts in Modern Science and Society.* London: Bloomsbury,
1989.

Brawley, Crista, and Erika Matunis. "Regeneration of male germline stem cells by spermatogonial dedifferentiation in vivo."
Science 304, no. 5675 (2004): 1331–34.

Briggs, Robert, and Thomas J. King. "Transplantation of living
nuclei from blastula cells into enucleated frogs' eggs." *Proceedings of the National Academy of Sciences* 38, no. 5 (1952): 455–63.

Brooks, William Keith. *The Law of Heredity: A Study of the Cause of
Variation, and the Origin of Living Organisms.* J. Murphy, 1883.

Browne, E. Janet. *Charles Darwin: The Power of Place*. Vol. 2. Princeton, NJ: Princeton University Press, 2003.

Burian, Richard M. "Challenges to the evolutionary synthesis." In *Evolutionary Biology*, edited by Bruce Wallace and Max K. Hecht, 247–69. Boston: Springer, 1988.

Buss, Leo W. *The Evolution of Individuality*. Princeton, NJ: Princeton University Press, 1987.

Cavazzana-Calvo, Marina, Salima Hacein-Bey, Geneviève de Saint Basile, Fabian Gross, Eric Yvon, Patrick Nusbaum, Françoise Selz, et al. "Gene therapy of human severe combined immuno-deficiency (SCID)-X1 disease." *Science* 288, no. 5466 (2000): 669–72.

Churchill, Frederick B. "August Weismann and a break from tradition." *Journal of the History of Biology* (1968): 91–112.

———. "Weismann's continuity of the germ-plasm in historical perspective." *Freiburger Universitätsblätter* 24, no. 87/88 (1985): 107–124.

———. "Weismann's Continuity of the Germ Plasm in Historical Perspective." In *August Weismann (1834–1914) und die theoretische Biologie des 19. Jahrhunderts. Urkunden, Berichte und Analysen*, edited by Klaus Sander, 107–130. Freiburg: Rombach, 1985.

———. "Weismann, Hydromedusae, and the biogenetic imperative: A reconsideration." In *A History of Embryology*, edited by T. Horder, J. A. Witkowski, and C. C. Wylie, 7–33. Cambridge: Cambridge University Press, 1986.

———. *August Weismann: Development, Heredity, and Evolution*. Cambridge, MA: Harvard University Press, 2015.

Cook-Deegan, Robert M. *The Gene Wars: Science, Politics, and the Human Genome*. W. W. Norton, 1994.

———. "Germ-line gene therapy: Keep the window open a crack."
Politics and the Life Sciences 13, no. 2 (1994): 217–20.

"CRISPR Everywhere" (editorial). *Nature News* 531 (2016): 155

Crowe, Nathan. *Forgotten Clones: The Birth of Cloning and the Biological Revolution.* University of Pittsburgh Press, 2021.

Dannenberg, Leah C., and Elaine C. Seaver. "Regeneration of the germline in the annelid Capitella teleta." *Developmental Biology* 440, no. 2 (2018): 74–87.

Darwin, Charles. *On the Origin of Species.* London: John Murray, 1859.

———. *The Variation of Animals and Plants under Domestication.* London: John Murray, 1868.

Davis, Robert L., Harold Weintraub, and Andrew B. Lassar. "Expression of a single transfected cDNA converts fibroblasts to myoblasts." *Cell* 51, no. 6 (1987): 987–1000.

De Vries, Hugo. *Intracellular Pangenesis.* Jena: Gustav Fischer, 1889.

Dietrich, Michael R. "Richard Goldschmidt's 'Heresies' and the Evolutionary Synthesis." *Journal of the History of Biology* 28, no. 3 (1995): 431–61.

do Nascimento Costa, José Jackson, Glaucinete Borges de Souza, Joyla Maria Pires Bernardo, Regislane Pinto Ribeiro, José Renato de Souza Passos, Francisco Taiã Gomes Bezerra, Márcia Viviane Alves Saraiva, and José Roberto Viana Silva. "Expression of markers for germ cells and oocytes in cow dermal fibroblast treated with 5-azacytidine and cultured in differentiation medium containing BMP2, BMP4 or follicular fluid." *Zygote* 25, no. 3 (2017): 341–57.

Dröscher, Ariane. "Images of cell trees, cell lines, and cell fates:

The legacy of Ernst Haeckel and August Weismann in stem cell research." *History and Philosophy of the Life Sciences* 36, no. 2 (2014): 157–86.

Dyce, Paul W., Lihua Wen, and Julang Li. "In vitro germline potential of stem cells derived from fetal porcine skin." *Nature Cell Biology* 8, no. 4 (2006): 384–90.

Dyce, Paul W., Wei Shen, Evanna Huynh, Hua Shao, Daniel AF Villagómez, Gerald M. Kidder, W. Allan King, and Julang Li. "Analysis of oocyte-like cells differentiated from porcine fetal skin-derived stem cells." *Stem Cells and Development* 20, no. 5 (2011): 809–19.

Eddy, E. M. "Germ plasm and the differentiation of the germ cell line." *International Review of Cytology* 43 (1976): 229–80.

Endersby, Jim. "Darwin on generation, pangenesis and sexual selection." In *The Cambridge Companion to Darwin*, edited by Jonathan Hodge and Gregory Radick, 69–91. Cambridge: Cambridge University Press, 2006.

Extavour, Cassandra G., and Michael Akam. "Mechanisms of germ cell specification across the metazoans: Epigenesis and preformation." *Development* (2003): 5869–84.

Frank, Uri, Günter Plickert, and Werner A. Müller. "Cnidarian interstitial cells: The dawn of stem cell research." In *Stem Cells in Marine Organisms*, edited by Baruch Rinkevich and Valeria Matranga, 33–59. Springer, Dordrecht, 2009.

Funayama, Noriko. "The stem cell system in demosponges: Insights into the origin of somatic stem cells." *Development, Growth & Differentiation* 52, no. 1 (2010): 1–14.

Gahan, James M., Brian Bradshaw, Hakima Flici, and Uri Frank. "The interstitial stem cells in Hydractinia and their role in regeneration." *Current Opinion in Genetics & Development* 40 (2016): 65–73.

Gao, Ming, and Alexey L. Arkov. "Next generation organelles: Structure and role of germ granules in the germline." *Molecular Reproduction and Development* 80, no. 8 (2013): 610–23.

Gartner, Anton, Peter R. Boag, and T. Keith Blackwell. "Germline survival and apoptosis." *WormBook: The Online Review of C. elegans Biology* (2018). Online at 10.1895/wormbook.1.145.1.

Geison, Gerald L. "Darwin and heredity: The evolution of his hypothesis of pangenesis." *Journal of the History of Medicine and Allied Sciences* 24, no. 4 (1969): 375–411.

Godfrey-Smith, Peter. *Darwinian Populations and Natural Selection.* Oxford University Press, 2009.

Gordon, Jon W., and Frank H. Ruddle. "Integration and stable germ line transmission of genes injected into mouse pronuclei." *Science* 214, no. 4526 (1981): 1244–46.

Griesemer, James R., and William C. Wimsatt. "Picturing Weismannism: A case study of conceptual evolution." In *What the Philosophy of Biology Is*, edited by Michael Ruse, 75–137. Springer, Dordrecht, 1989.

Gurchot, Charles. "*L'origine des cellules reproductrices et le problème de la lignée germinal* and *Continuité germinale et reproduction agame.* L Bounoure. Paris, France: Gauthiers-Villars, 1939; 1940. Pp. xii + 271; pp. 83. (Illustrated)." Review. *Science* 105 (1947): 106–7.

Gurdon, John B. "The developmental capacity of nuclei taken from intestinal epithelium cells of feeding tadpoles." *Journal of Experimental Morphology* 10 (1962): 622–40.

Gurdon, John B., and James A. Byrne. "The history of cloning." In

Ethical Eye: Cloning, edited by Anne McLaren, 35–54. Germany: Council of Europe Publishers, 2002.

Gurdon, John B., and James A. Byrne. "The first half-century of nuclear transplantation." *Proceedings of the National Academy of Sciences* 100, no. 14 (2003): 8048–52.

Han, Yanyan, Dandan Liu, and Lianhong Li. "PD-1/PD-L1 pathway: Current researches in cancer." *American Journal of Cancer Research* 10, no. 3 (2020): 727–42.

Hargitt, George T. "Germ cells of Coelenterates. I. *Campanularia flexuosa.*" *Journal of Morphology* 24 (1913): 383–420.

———. "Germ cells of Coelenterates. II. *Clava leptostyla.*" *Journal of Morphology* 27 (1916): 85–98.

———. "Germ cells of Coelenterates. III. *Aglanthia digitalis.* IV. *Hybocadon prolifer.*" *Journal of Morphology* 28 (1917): 593–642.

———. "Germ cells of Coelenterates. V. *Eudendrium ramosum.*" *Journal of Morphology* 31 (1918): 1–24.

———. "Germ cells of Coelenterates. VI. General considerations, discussion, conclusions." *Journal of Morphology* 33 (1919): 1–60.

———. "Germ cell origins in the adult salamander, *Diemyctylus viridescens.*" *Journal of Morphology and Physiology* 39 (1924): 63–111.

———. "The formation of the sex glands and germ cells of mammals. 1. The origin of the germ cells in the albino rat." *Journal of Morphology and Physiology* 40 (1925): 517–57.

———. "The formation of the sex glands and germ cells of mammals. 2. The history of the male germ cells in the albino rat." *Journal of Morphology and Physiology* 42 (1926): 253–305.

———. "What is germ plasm?" *Science* 100 (1944): 343–45.

Hayashi, Katsuhiko, Hiroshi Ohta, Kazuki Kurimoto, Shinya Aramaki, and Mitinori Saitou, "Reconstitution of the mouse germ cell specification pathway in culture by pluripotent stem cells." *Cell* 146, no. 4 (2011): 519–32.

Hayashi, Katsuhiko, Sugako Ogushi, Kazuki Kurimoto, So Shimamoto, Hiroshi Ohta, and Mitinori Saitou. "Offspring from oocytes derived from in vitro primordial germ cell–like cells in mice." *Science* 338, no. 6109 (2012): 971–75.

Hegner, Robert W. "Effects of removing the germ-cell determinants from the eggs of some chrysomelid beetles. Preliminary report." *Biological Bulletin* 16, no. 1 (1908): 19–26.

———. "The origin and early history of the germ cells in some Chrysomelid beetles." *Journal of Morphology* 20 (1909): 231–95.

———. "Germ-cell determinants and their significance." *American Naturalist* 45 (1911): 385–97.

———. "The germ cell determinants in the eggs of Chrysomelid beetles." *Science* 33 (1911): 71–72.

———. *The Germ-Cell Cycle in Animals.* New York: Macmillan, 1914.

———. "Studies on germ cells." *Journal of Morphology* 25 (1914): 375–509.

Hendriks, S., N. A. A. Giesbertz, A. L. Bredenoord, and S. Repping. "Reasons for being in favour of or against genome modification: A survey of the Dutch general public." *Human Reproduction Open* 2018, no. 3 (2018): 1–12.

Heys, Florence. "The problem of the origin of germ cells." *Quarterly Review of Biology* 6, no. 1 (1931): 1–45.

Hikabe, Orie, Nobuhiko Hamazaki, Go Nagamatsu, Yayoi Obata, Yuji Hirao, Norio Hamada, So Shimamoto, et al. "Reconstitu-

tion in vitro of the entire cycle of the mouse female germ line."
Nature 539, no. 7628 (2016): 299–303.

Holterhoff, Kate. "The history and reception of Charles Darwin's hypothesis of pangenesis." *Journal of the History of Biology* 47, no. 4 (2014): 661–95.

Hughes, Sally Smith. "Making dollars out of DNA: The first major patent in biotechnology and the commercialization of molecular biology, 1974–1980." *Isis* 92, no. 3 (2001): 541–75.

Hurlbut, J. Benjamin. "Limits of responsibility: Genome editing, Asilomar, and the politics of deliberation." *Hastings Center Report* 45, no. 5 (2015): 11–14.

———. *Experiments in Democracy.* New York: Columbia University Press, 2017.

———. "Human genome editing: Ask whether, not how." *Nature* 565, no. 7738 (2019): 135–36.

Huxley, Julian. "Some war-time biological books in French." *Science* 157 (1946): 611–14.

Irie, Naoko, Leehee Weinberger, Walfred W. C. Tang, Toshihiro Kobayashi, Sergey Viukov, Yair S. Manor, Sabine Dietmann, Jacob H. Hanna, and M. Azim Surani. "SOX17 is a critical specifier of human primordial germ cell fate." *Cell* 160, no. 1–2 (2015): 253–68.

Jasanoff, Sheila, J. Benjamin Hurlbut, and Krishanu Saha. "CRISPR democracy: Gene editing and the need for inclusive deliberation." *Issues in Science and Technology* 32, no. 1 (2015): 25–32.

Jinek, Martin, Krzysztof Chylinski, Ines Fonfara, Michael Hauer, Jennifer A. Doudna, and Emmanuelle Charpentier. "A programmable dual-RNA–guided DNA endonuclease in adaptive bacterial immunity." *Science* 337, no. 6096 (2012): 816–21.

Johnson, Joshua, Jacqueline Canning, Tomoko Kaneko, James K. Pru, and Jonathan L. Tilly. "Germline stem cells and follicular renewal in the postnatal mammalian ovary." *Nature* 428, no. 6979 (2004): 145–50.

Johnson, Joshua, Jessamyn Bagley, Malgorzata Skaznik-Wikiel, Ho-Joon Lee, Gregor B. Adams, Yuichi Niikura, Katherine S. Tschudy, et al. "Oocyte generation in adult mammalian ovaries by putative germ cells in bone marrow and peripheral blood." *Cell* 122, no. 2 (2005): 303–15.

Kaczmarczyk, Angela N. "Germline maintenance and regeneration in the amphipod crustacean, *Parhyale hawaiensis*." PhD diss. UC Berkeley, 2014.

Kai, Toshie, and Allan Spradling. "Differentiating germ cells can revert into functional stem cells in Drosophila melanogaster ovaries." *Nature* 428, no. 6982 (2004): 564–69.

Kerr, J. B., and K. E. Dixon. "An ultrastructural study of germ plasm in spermatogenesis of Xenopus loevis." *Journal of Embryology and Experimental Morphology* 32 (1974): 573–92.

King, Thomas J., and Robert Briggs. "Transplantation of living nuclei of late gastrulae into enucleated eggs of Rana pipiens." *Journal of Experimental Morphology* 2 (1954): 73–80.

Kohler, Robert E. *Lords of the fly: Drosophila genetics and the experimental life.* Chicago: University of Chicago Press, 1994.

Kohn, Donald B. "Gene therapy for XSCID: The first success of gene therapy." *Pediatric Research* 48, no. 5 (2000): 578.

Lamarck, Jean-Baptiste. *Philosophie zoologique ou exposition des* considérations relatives à l'histoire naturelle des animaux. Paris: Musée d'Histoire Naturelle, 1809.

Lanfear, Robert. "Do plants have a segregated germline?" PLoS Biology 16, no. 5 (2018): e2005439.

Laplane, Lucie. Cancer Stem Cells: Philosophy and Therapies. Cambridge, MA: Harvard University Press, 2016.

Lázár-Molnár, Eszter, Bing Chen, Kari A. Sweeney, Emilie J. Wang, Weijun Liu, Juan Lin, Steven A. Porcelli, Steven C. Almo, Stanley G. Nathenson, and William R. Jacobs Jr. "Programmed death-1 (PD-1)–deficient mice are extraordinarily sensitive to tuberculosis." Proceedings of the National Academy of Sciences 107, no. 30 (2010): 13402–407.

Leclère Lucas, Muriel Jager, Carine Barreau, Patrick Chang, Hervé Le Guyader, Michael Manuel, and Evelyn Houliston. "Maternally localized germ plasm mRNAs and germ cell/stem cell formation in the cnidarian Clytia." Developmental Biology 364, no. 2 (2012): 236–48.

Ledford, Heidi. "CRISPR, the disruptor." Nature 522, no. 7544 (2015): 20–25.

Leeuwenhoek, Antonie van. "Observationes D. Anthonii Lewenhoeck, De Natis E Semine Genitali Animalculis." Philosophical Transactions 12 (1678): 1040–46.

Leuckart, Rudolf. "Die ungeschlechtliche Fortpflanzung der Cecidomyienlarven." Archiv für Naturgeschichte 31, no. 1 (1865): 286–303.

Lewens, Tim. "Blurring the germline: Genome editing and transgenerational epigenetic inheritance." Bioethics 34, no. 1 (2020): 7–15.

Liang, Puping, Yanwen Xu, Xiya Zhang, Chenhui Ding, Rui Huang, Zhen Zhang, Jie Lv, et al. "CRISPR/Cas9-mediated gene editing in human tripronuclear zygotes." *Protein & Cell 6,* no. 5 (2015): 363–72.

MacCord, Kate, and B. D. Özpolat. "Is the germline immortal and continuous? A discussion in light of iPSCs and germline regeneration" (2019). Online at https://doi.org/10.5281/zenodo.3385322.

Maienschein, Jane. "Cell lineage, ancestral reminiscence, and the biogenetic law." *Journal of the History of Biology* (1978): 129–58.

———. "Shifting assumptions in American biology: Embryology, 1890–1910." *Journal of the History of Biology* (1981): 89–113.

———. *Transforming Traditions in American Biology, 1880–1915.* Baltimore: Johns Hopkins University Press, 1991.

———. "From presentation to representation in EB Wilson's *The Cell.*" *Biology and Philosophy* 6, no. 2 (1991): 227–54.

———. "On cloning: Advocating history of biology in the public interest." *Journal of the History of Biology* 34, no. 3 (2001): 423–32.

———. "Regenerative medicine's historical roots in regeneration, transplantation, and translation." *Developmental Biology* 358, no. 2 (2011): 278–84.

———. *Whose View of Life?* Cambridge, MA: Harvard University Press, 2005.

Maienschein, Jane, and Kate MacCord. *What Is Regeneration?* Chicago: University of Chicago Press, 2022.

Makar, Karen, and Kotaro Sasaki. "Roadmap of germline development and in vitro gametogenesis from pluripotent stem cells." *Andrology* 8, no. 4 (2020): 842–51.

Marchione, Marilyn. "Chinese researcher claims first gene-edited babies." 26 November 2018. https://www.apnews.com/4997bb 7aa36c45449b488e19ac83e86d.

Mayr, Ernst. *The Growth of Biological Thought: Diversity, Evolution, and Inheritance.* Cambridge, MA: Harvard University Press, 1982.

Mayr, Ernst, and William B. Provine, eds. *The Evolutionary Synthesis: Perspectives on the Unification of Biology.* Cambridge, MA: Harvard University Press, 1998.

McCaughey, Tristan, Paul G. Sanfilippo, George EC Gooden, David M. Budden, Li Fan, Eva Fenwick, Gwyneth Rees, et al. "A global social media survey of attitudes to human genome editing." *Cell Stem Cell* 18, no. 5 (2016): 569–72.

Mendel, Gregor. "Versuche über pflanzen-hybriden." *Verhandlungen des naturforschenden Vereins in Brünn* 4 (1866): 3–47.

Metschnikoff, Élie. "Ueber die Entwicklung der Cecidomyienlarven aus dem Pseudovum." *Archiv für Naturgeschichte* 31, no. 1 (1865): 304–10. (Note that Metschnikoff is spelled "Mecznikoff" in the journal.)

Modrell, Melinda Sue. "Early cells fate specification in the amphipod crustacean, *Parhyale hawaiensis.*" PhD diss. University of California, Berkeley, 2007.

Morgan, Thomas Hunt. "Growth and regeneration in *Planaria lugubris.*" *Archiv für Entwicklungsmechanik der Organismen* 13 (1901): 179–212.

Morgan, Thomas Hunt, Alfred Henry Sturtevant, Hermann Joseph Muller, and Calvin Blackman Bridges. *The Mechanism of Mendelian Heredity.* H. Holt, 1915.

Müller, Werner A., Regina Teo, and Uri Frank. "Totipotent migratory stem cells in a hydroid." *Developmental Biology* 275, no. 1 (2004): 215–24.

Mussolino, Claudio, and Toni Cathomen. "On target? Tracing zinc-finger-nuclease specificity." *Nature Methods* 8, no. 9 (2011): 725–26.

National Academies of Sciences, Engineering, and Medicine. *International Summit on Human Gene Editing: A Global Discussion.* Washington, DC: National Academies Press, 2016.

———. *Human Genome Editing: Science, Ethics, and Governance.* Washington, DC: National Academies Press, 2017.

———. "Second international summit on human genome editing. Continuing the global discussion: Proceedings of a workshop—in brief." Washington, DC: National Academies Press, 2019.

———. *Heritable Human Genome Editing.* Washington, DC: National Academies Press, 2020.

Newmark, P. A., Y. Wang, and T. Chong. "Germ cell specification and regeneration in planarians." *Cold Spring Harbor Symposia on Quantitative Biology* 73 (2008): 573–81.

Nishimura, H., M. Nose, H. Hiai, N. Minato, and T. Honjo. "Development of lupus-like autoimmune diseases by disruption of the PD-1 gene encoding an ITIM motif-carrying immunoreceptor." *Immunity* 11, no. 2 (1999): 141–51.

Nishimura, Hiroyuki, Taku Okazaki, Yoshimasa Tanaka, Kazuki Nakatani, Masatake Hara, Akira Matsumori, Shigetake Sasayama, et al. "Autoimmune dilated cardiomyopathy in PD-1 receptor-deficient mice." *Science* 291, no. 5502 (2001): 319–22.

Okazaki, Taku, Yoshimasa Tanaka, Ryosuke Nishio, Tamotsu Mitsuiye, Akira Mizoguchi, Jian Wang, Masayoshi Ishida, et al.

"Autoantibodies against cardiac troponin I are responsible for dilated cardiomyopathy in PD-1-deficient mice." *Nature Medicine* 9, no. 12 (2003): 1477–83.

Okita, Keisuke, Tomoko Ichisaka, and Shinya Yamanaka. "Generation of germline-competent induced pluripotent stem cells." *Nature* 448, no. 7151 (2007): 313–17.

Olby, Robert C. "Charles Darwin's manuscript of pangenesis." *British Journal for the History of Science* 1, no. 3 (1963): 251–63.

———. "The dimensions of scientific controversy: The biometric-Mendelian debate." *British Journal for the History of Science* 22, no. 3 (1989): 299–320.

———. "The emergence of genetics." In *Companion to the History of Modern Science*, edited by G. N. Cantor, J. R. R. Christie, M. J. S. Hodge, and R. C. Olby, 521–36. New York: Routledge, 1996.

Özpolat, B. Duygu, Emily S. Sloane, Eduardo E. Zattara, and Alexandra E. Bely. "Plasticity and regeneration of gonads in the annelid *Pristina leidyi.*" *EvoDevo* 7, no. 1 (2016): 1–15.

Perry, Mary Ellen, Kayla M. Valdes, Elizabeth Wilder, Christopher P. Austin, and Philip J. Brooks. "Genome editing to 're-write' wrongs." *Nature Review Drug Discovery* 17, no. 10 (2018): 689–90.

President's Commission for the Study of Ethical Problems in Medicine and Biomedical and Behavioral Research (Morris B. Abram, chairman). *Splicing Life: A Report on the Social and Ethical Issues of Genetic Engineering with Human Beings.* President's Commission for the Study of Ethical Problems in Medicine and Biomedical and Behavioral Research, 1982.

Provine, William B. *The Origins of Theoretical Population Genetics.* Chicago: University of Chicago Press, 1971.

Rabino, Isaac. "The impact of activist pressures on recombinant DNA research." *Science, Technology, & Human Values* 16, no. 1 (1991): 70–87.

Raper, Steven E., Narendra Chirmule, Frank S. Lee, Nelson A. Wivel, Adam Bagg, Guang-ping Gao, James M. Wilson, and Mark L. Batshaw. "Fatal systemic inflammatory response syndrome in a ornithine transcarbamylase deficient patient following adenoviral gene transfer." *Molecular Genetics and Metabolism* 80, nos. 1–2 (2003): 148–58.

Richmond, Marsha L. "Women in the early history of genetics: William Bateson and the Newnham College Mendelians, 1900–1910." *Isis* 92, no. 1 (2001): 55–90.

Robinson, Gloria. *A Prelude to Genetics: Theories of a Material Substance of Heredity—Darwin to Weismann.* Lawrence, KS: Coronado Press, 1979.

———. "August Weismann's hereditary theory." In *August Weismann (1834–1914) und die theoretische Biologie des 19. Jahrhunderts. Urkunden, Berichte und Analysen,* edited by Klaus Sander, 83–90. Freiburg: Rombach, 1985.

Rubaschkin, W. "Über die Urgeschlechtszellen bei Säugetieren." *Anatomische Hefte* 39 (1910): 603–52.

———. "Zur Lehre von der Keimbahn bei Säugetieren. Über die Entwickelung der Keimdrüsen." *Anatomische Hefte* 46 (1912): 343–411.

Sarton, George. "The discovery of the mammalian egg and the foundation of modern embryology." *Isis* 16, no. 2 (1931): 315–77.

Sasaki, Kotaro, Shihori Yokobayashi, Tomonori Nakamura, Ikuhiro Okamoto, Yukihiro Yabuta, Kazuki Kurimoto, Hiroshi Ohta, et al. "Robust in vitro induction of human germ cell fate from pluripotent stem cells." *Cell Stem Cell* 17, no. 2 (2015): 178–94.

Sawada, Hitoshi, Masaya Morita, and Megumi Iwano. "Self/ non-self recognition mechanisms in sexual reproduction: new insight into the self-incompatibility system shared by flowering plants and hermaphroditic animals." *Biochemical and Biophysical Research Communications* 450, no. 3 (2014): 1142–48.

Scheider, Karl Camillo. "Histologie von *Hydra* fusca mit besonderer Berücksichtigung des Nervensystems der Hydropolypen." *Archiv für mikroskopische Anatomie* 35 (1890): 321–79.

Scheufele, Dietram A., Michael A. Xenos, Emily L. Howell, Kathleen M. Rose, Dominique Brossard, and Bruce W. Hardy. "US attitudes on human genome editing." *Science* 357, no. 6351 (2017): 553–54.

Schumann, Kathrin, Steven Lin, Eric Boyer, Dimitre R. Simeonov, Meena Subramaniam, Rachel E. Gate, Genevieve E. Haliburton, et al. "Generation of knock-in primary human T cells using Cas9 ribonucleoproteins." *Proceedings of the National Academy of Sciences* 112, no. 33 (2015): 10437–442.

Schwank, Gerald, Bon-Kyoung Koo, Valentina Sasselli, Johanna F. Dekkers, Inha Heo, Turan Demircan, Nobuo Sasaki, et al. "Functional repair of CFTR by CRISPR/Cas9 in intestinal stem cell organoids of cystic fibrosis patients." *Cell Stem Cell* 13, no. 6 (2013): 653–58.

Siebert, Stefan, Friederike Anton-Erxleben, and Thomas C. G. Bosch. "Cell type complexity in the basal metazoan *Hydra* is maintained by both stem cell based mechanisms and trans-differentiation." *Developmental Biology* 313, no. 1 (2008): 13–24.

Simkins, Cleveland Sylvester. "On the origin and migration of the so-called Primordial Germ Cells in the mouse and the rat." *Acta Zoologica* 4, no. 2–3 (1923): 241–84.

Simpson, George Gaylord, Colin S. Pittendrigh, and Lewis H. Tiffany. *Life: An Introduction to Biology.* New York: Harcourt, Brace and Company, 1957.

Smocovitis, Vassiliki Betty. "Unifying biology: The evolutionary synthesis and evolutionary biology." *Journal of the History of Biology* 25, no. 1 (1992): 1–65.

Sober, Elliott R. "The Modern Synthesis: Its scope and limits." *PSA: Proceedings of the Biennial Meeting of the Philosophy of Science Association* 1982 (1982): 314–21.

Solana, Jordi. "Closing the circle of germline and stem cells: The Primordial Stem Cell hypothesis." *EvoDevo* 4, no. 1 (2013): 1–17.

Stadtfeld, Matthias, and Konrad Hochedlinger. "Induced pluripotency: History, mechanisms, and applications." *Genes & Development* 24, no. 20 (2010): 2239–63.

Stanford, P. Kyle. "Darwin's pangenesis and the problem of unconceived alternatives." *British Journal for the Philosophy of Science* 57, no. 1 (2006): 121–44.

Su, Shu, Bian Hu, Jie Shao, Bin Shen, Juan Du, Yinan Du, Jiankui Zhou, et al. "CRISPR-Cas9 mediated efficient PD-1 disruption on human primary T cells from cancer patients." *Scientific Reports* 6 (2016): 20070.

Takahashi, Kazutoshi, and Shinya Yamanaka. "Induction of pluripotent stem cells from mouse embryonic and adult fibroblast cultures by defined factors." *Cell* 126, no. 4 (2006): 663–76.

Takahashi, Kazutoshi, Koji Tanabe, Mari Ohnuki, Megumi Narita,

Tomoko Ichisaka, Kiichiro Tomoda, and Shinya Yamanaka. "Induction of pluripotent stem cells from adult human fibroblasts by defined factors." *Cell* 131, no. 5 (2007): 861–72.

Takamura, Katsumi, Miyuki Fujimura, and Yasunori Yamaguchi. "Primordial germ cells originate from the endodermal strand cells in the ascidian *Ciona intestinalis.*" *Development Genes and Evolution* 212, no. 1 (2002): 11–18.

Tesson, Laurent, Claire Usal, Séverine Ménoret, Elo Leung, Brett J. Niles, Séverine Remy, Yolanda Santiago, et al. "Knockout rats generated by embryo microinjection of TALENs." *Nature Biotechnology* 29, no. 8 (2011): 695–96.

Travis, John. 2015. "Breakthrough of the year: CRISPR makes the cut." *Science Magazine*: 1456–1457.

Van Beneden, Edouard. "Recherches sur la composition et la signification de l'œuf: basées sur l'étude de son mode de formation et des premiers phénomènes embryonnaires (mammifères, oiseaux, crustacés, vers)." *L'Academie royale des sciences de Beligique* 34: 1–238 (presented 1868; published 1870).

van Wolfswinkel, Josien C., Daniel E. Wagner, and Peter W. Reddien. "Single-cell analysis reveals functionally distinct classes within the planarian stem cell compartment." *Cell Stem Cell* 15, no. 3 (2014): 326–39.

Voronina, Ekaterina, Manuel Lopez, Celina E. Juliano, Eric Gustafson, Jia L. Song, Cassandra Extavour, Sophie George, Paola Oliveri, David McClay, and Gary Wessel. "Vasa protein expression is restricted to the small micromeres of the sea urchin, but is inducible in other lineages early in development." *Developmental Biology* 314, no. 2 (2008): 276–86.

Waldeyer, Heinrich Wilhelm Gottfried. *Eierstock und Ei: ein Beitrag zur Anatomie und Entwicklungeschichte der Sexualorgane.* Leipzig: Wilhelm Engelmann, 1870.

Wang, Jian, Taku Yoshida, Fumio Nakaki, Hiroshi Hiai, Taku Okazaki, and Tasuku Honjo. "Establishment of NOD-Pdcd1-/-mice as an efficient animal model of type I diabetes." *Proceedings of the National Academy of Sciences* 102, no. 33 (2005): 11823–828.

Wang, Jiang-Hui, Rong Wang, Jia Hui Lee, Tiara WU Iao, Xiao Hu, Yu-Meng Wang, Lei-Lei Tu, et al. "Public attitudes toward gene therapy in China." *Molecular Therapy-Methods & Clinical Development* 6 (2017): 40–42.

Weismann, August. "Die Entwicklung der Dipteren im Ei." *Zeitschrift zur Wissenschaft Zoologie* 13 (1863): 107–220.

———. *Die Entstehung der Sexualzellen bei den Hydromedusen.* Berlin: Fischer, 1883

———. *Essays upon Heredity and Kindred Biological Problems.* Translated by Edward Poulton, Selmar Schönland, and Arthur Shipley. Oxford: Clarendon Press, 1889.

———. *The Germ Plasm.* English translation by W. N. Parker and H. Ronnfeldt. New York: Scribners, 1892.

White, Yvonne A. R., Dori C. Woods, Yasushi Takai, Osamu Ishihara, Hiroyuki Seki, and Jonathan L. Tilly. "Oocyte formation by mitotically active germ cells purified from ovaries of reproductive-age women." *Nature Medicine* 18, no. 3 (2012): 413–21.

Whitman, Charles Otis. *The Embryology of Clepsine.* J. E. Adlard, 1878.

Wilson, Edmund Beecher. "The cell lineage of Nereis: A contribution to the cytogeny of the annelid body." *Journal of Morphology* 6, no. 3: 361–480.

———. *The Cell in Development and Inheritance*. London: Macmillan and Co., 1896.

Winther, Rasmus G. "Darwin on variation and heredity." *Journal of the History of Biology* 33, no. 3 (2000): 425–55.

Wirth, Thomas, Nigel Parker, and Seppo Ylä-Herttuala. "History of gene therapy." *Gene* 525, no. 2 (2013): 162–69.

World Health Organization. "Statement on governance and oversight of human genome editing." 26 July 2019. https://www.who.int/news/item/26-07-2019-statement-on-governance-and-oversight-of-human-genome-editing.

Wright, Susan. "Recombinant DNA technology and its social transformation, 1972–1982." *Osiris* 2 (1986): 303–60.

Yajima, Mamiko, and Gary M. Wessel. "Small micromeres contribute to the germline in the sea urchin." *Development* 138, no. 2 (2011): 237–43.

Yamashita, Grant. "On the germ-soma distinction in evolutionary biology: A historical and conceptual approach." PhD diss. University of California, Davis, 2006.

Yamashita, Yukiko. "Unsolved problems in cell biology: Germline immortality." *ASCB Newsletter* 42 no. 1 (2019): 15–16

Yamashiro, Chika, Kotaro Sasaki, Yukihiro Yabuta, Yoji Kojima, Tomonori Nakamura, Ikuhiro Okamoto, Shihori Yokobayashi, et al. "Generation of human oogonia from induced pluripotent stem cells in vitro." *Science* 362, no. 6412 (2018): 356–60.

Yi, Doogab. "Who owns what? Private ownership and the public interest in recombinant DNA technology in the 1970s." *Isis* 102, no. 3 (2011): 446–74.

Yoshida, Keita, Akiko Hozumi, Nicholas Treen, Tetsushi Sakuma, Takashi Yamamoto, Maki Shirae-Kurabayashi, and Yasunori

Sasakura. "Germ cell regeneration-mediated, enhanced mutagenesis in the ascidian *Ciona intestinalis* reveals flexible germ
cell formation from different somatic cells." *Developmental Biology* 423, no. 2 (2017): 111–25.

Yu, Junying, Maxim A. Vodyanik, Kim Smuga-Otto, Jessica
Antosiewicz-Bourget, Jennifer L. Frane, Shulan Tian, Jeff Nie,
et al. "Induced pluripotent stem cell lines derived from human
somatic cells." *Science* 318, no. 5858 (2007): 1917–20.

Zhao, Ruifeng, Qisheng Zuo, Xia Yuan, Kai Jin, Jing Jin, Ying Ding,
Chen Zhang, et al. "Production of viable chicken by allogeneic
transplantation of primordial germ cells induced from somatic
cells." *Nature Communications* 12, no. 1 (2021): 1–13.

Index

Page numbers in italics refer to figures.

acquired characteristics, heredity theories, 15–18, 25, 35
ADA SCID disorder, 111–12
annelids, natural transdifferentiation, 89–91
archeocyte cells, sponges, 79–82
assumptions, as science element, 9–10. *See also* continuity theories; identifying germ cells; Weismann Barrier

babies, genetically modified, 116–17
Baer, Karl Ernst von, 12, 129n3
Bateson, William, 35–36
Berrill, John Normal, 34–35, 58–59
ß-thalassemia, 115–16
biophors, in Weismann's germ plasm architecture, 20
blastema, defined, 5
Bounoure, Louis, 55–58, 66–67
Briggs, Robert, 95–96
Brooks, William Keith, 17

cancers, gene therapies, 115, 119–20
Capitella teleta, natural transdifferentiation, 89–90
Cell in Development and Inheritance, The (Wilson), 27–28, 37, 130n25
cell lineage, defined, 2–3
cell-lineage tracing, historical studies, 30–34, 49–50, 64–65. *See also* continuity theories; identifying germ cells
cell reprogramming, historical perspectives, 95–97. *See also* induced transdifferentiation; natural transdifferentiation
cell theory, Wilson's approach, 28–29
Charpentier, Emmanuelle, 111
choanocyte cells, sponges, 79–82
chromosomes, 19–20, 36, 37. *See also* gene expression methods; gene therapies; genetics research; genome editing
Churchill, Frederick B., 17, 27
Ciona intestinalis, natural transdifferentiation, 85–89
Clepsine studies, Whitman's, 32
clustered regularly interspaced short palindromic repeats (CRISPR), 110–13
continuity theories: Bounoure's claims, 55–58; and differentiation modes, 68–70; genetics vs. embryology perspectives, 34–35; Hargitt's criticisms, 44–45, 46, 50; Hegner's studies, 53–55, 66–67; inter-generation problem, 68, 70–71; intra-organism problem, 68, 71–72; limitations summarized, 58–60, 65–75, 104–6; in Weismann's germ plasm theory, 25–26, 30; Wilson's representation of, 30

Correns, Carl, 35
CRISPR (clustered regularly interspaced
    short palindromic repeats), 110–13
crustaceans, natural transdifferentiation,
    91–93
cytoplasmic granules, 51–52

Dannenberg, Leah, 89–90, 98–99
Darwin, Charles, 14–18
Davis, Robert, 96
dedifferentiated, defined, 5
determinants, in Weismann's germ plasm
    architecture, 20, 22
de Vries, Hugo, 17, 35
Diemyctylus viridescens experiments,
    Hargitt's, 45–46
differentiation processes, overview, 68–70.
    See also continuity theories; germ cells
    and related entries; induced trans-
    differentiation; *and other specific topics*
differentiation property, stem cells, 76–78
Dixon, Keith E., 67
Doudna, Jennifer, 111
Driesch, Hans, 95
Drosophila melanogaster (fruit flies), 67,
    68–69, 74, 133n2

echinoderms, natural transdifferentiation,
    93–94
Eddy, Edward Mitchell, 59–60
empirical vs. normative claim, definitions,
    42
epigenesis, 69–70, 71, 72, 94, 105
ethics debates, genome editing, 114–18
evolution theories, nineteenth century,
    14–18. *See also* Weismann, August;
    Weismann Barrier
extraembryonic ectoderm, mouse
    embryos, 101

family tree comparison, 2–3
fibroblasts-to-myoblasts, 96
frog studies, 55–56, 95–96
fruit flies, 67, 68–69, 74, 133n2

gametes, defined, 2
gametes, discovery origins, 11–14

gastrulation, 70
g-cell, *Parhyale hawaiensis*, 92
gemmules, in pangenesis theory, 15–17
gene expression methods, 63–64
gene therapies, 111–13, 119–22
genetics research: focus on heredity prob-
    lem, 34–38; naming conventions, 133n2;
    role of Weismann Barrier, 33–41
genome editing: defined, 109; and gene
    therapies, 111–13; implications of
    unchallenged historic assumptions,
    118–24; philosophy's role, 123–24;
    policies and ethics debates, 114–18;
    technologies for, 109–11
germ cell continuity. *See* continuity
    theories
Germ-Cell Cycle in Animals, The (Hegner),
    53
germ cell determinants, Hegner's, 51–52
germ cell–like cells, 100
germ cell lineage. *See* continuity theories;
    germline regeneration, overviews;
    germline regeneration, traditional
    perspectives
germ cells: defined, 1, 2; differentiation
    processes summarized, 68–70;
    genome editing debates, 114–22;
    identification assumption, 62–65,
    103–4. *See also* germline regeneration,
    overviews
germ cells, from metazoan somatic cells:
    in annelids, 89–90; in mice, 99; in
    Morgan's planarian experiments, 84;
    overview, 103; in tunicates, 88–89.
    See also induced transdifferentiation;
    natural transdifferentiation
germ cells, historic theorizing: Hargitt's
    experiments, 44–50, 62–63; Hegner's
    studies, 50–55; in Simpson's biology
    textbook, 37–39; in Weismann's germ
    plasm theorizing, 22, 23, 24–25, 52–54;
    in Wilson's cell theorizing, 28, 29–30, 37
germ layers, in embryological studies,
    32, 46
germline regeneration, overviews: benefits
    of expanded perspective, 106–7; defi-
    nitions, 2, 68; elements defined, 2–4;

genetics vs. embryology questions,
34–35; initiation, 3; models of, 73–76
germline regeneration, traditional perspectives: benefits of historical/philosoph-
ical investigation, 6–7, 123–24; contra-
dictory evidence summarized, 104–6;
historic assumptions summarized,
3–7, 61–62, 103–4; research repercussions, 106–9. *See also* continuity
theories; genome editing; Weismann,
August; Weismann Barrier
Germ Plasm, The (Weismann), 19, 130n25
germ plasm continuity. *See* continuity theories; Weismann, August; Weismann
Barrier
germ-track, in Weismann's germ plasm
architecture, 22–25, 30, 52
Growth of Biological Thought, The (Mayr),
39–40
Gurdon, John, 95–96, 97

Hargitt, George T., 44–50, 56, 62–63
Hayashi, Katsunhiko, 99–101
He, Jiankui, 116–17
Hegner, Robert, 50–58, 66–67
heredity theories: with genetics studies,
33–38; nineteenth century, 14–18. *See
also* Weismann, August; Weismann
Barrier
heterokinesis, in Weismann's germ plasm
theory, 21–22, 26
Heys, Florence, 43–44
homeokinesis, in Weismann's germ plasm
theory, 21–22
HTT gene, alteration debate, 115
human cells, transdifferentiation studies,
97–99, 101–3
Human Genome Editing, 117–18
Huntington's disease, 115
Huxley, Julian, 57
Hydra, transdifferentiation, 85
Hydractinia, 82–83
Hydrozoans, 19, 44–45, 82–83

i-cells, *Hydractinia*, 82–83
idants, in Weismann's germ plasm architecture, 20

identifying germ cells: Hargitt's arguments
about accuracy, 44, 47–50, 56, 62–63;
historic assumption, 5–6, 62, 103–4;
methodological challenges, 62–65, 104
idioplasm, in Weismann's germ plasm
architecture, 22, 25–26
ids, in Weismann's germ plasm architecture, 20
induced pluripotent stem cells (iPSCs),
97, 99–103
induced transdifferentiation: evidence
standard, 98–99; in humans, 101–3; in
mice, 98–101; research history, 94–97
interstitial cells, *Hydractinia*, 82–83
iPSCs (induced pluripotent stem cells),
97, 99–103

Jaeger, Gustav, 23–24, 68, 70, 130n19
Johannsen, Wilhelm, 35

keimbahn-determinant, Hegner's concept,
52–54
Kerr, Jeffrey B., 67
King, Thomas, 95–96

Lamarck, Jean-Baptiste, 15
Lassar, Andrew, 96
leaf beetles, Hegner's studies, 50–51
leech egg studies, Whitman's, 32
Leeuwenhoek, Antonie van, 111
leukemia, 113
Life (Simpson), 38
Liu, Chien-Kang, 34–35, 58–59

Maienschein, Jane, 32
maternal inheritance (preformation), 68–
69, 72, 92
Mayr, Ernst, 39–40, 57
Mechanism of Mendelian Heredity, The
(Morgan), 36
Mendel, Gregor, 35
Metschnikoff, Élie, 52
mice, 4–5, 69–70, 96–101, 102, 120
micromere cells, sea urchins, 94
Modern Synthesis movement, 39–40, 57
Modrell, Melinda, 92
Morgan, Thomas Hunt, 36, 84, 131n35

morphological criteria problem, germ
cells, 47–48
Müller, Werner, 83
multipotency ability, stem cells, 64, 76–77
myoblasts-from-fibroblasts, 96

natural transdifferentiation: in annelids,
89–91; in crustaceans, 91–93; overview, 85; in tunicates, 85–89
neoblasts, 84
newt experiments, Hargitt's, 45–46
normative vs. empirical claim, definitions,
42
nuclear division processes, Weismann's
theorizing, 21–22, 25–26
Nussbaum, Moritz, 23–24, 68, 70

oocytes, 2, 68–69, 81, 95, 100–101
ova, documentation history, 12–13
Özpolat, B. Duygu, 91, 134n10

pangenesis theory, Darwin's, 15–17, 25
Parhyale hawaiensis, natural transdifferentiation, 92–93
PD-1 gene, editing implications, 119–20
phylogeny of metazoans, 77, *80, 86, 98*
Piwi, defined, 133n2
planarians, 83–84
pluripotency model, germline regeneration, 74, 75, 76–85
pole cells, in Hegner's leaf beetle studies,
50–51
pole-disc granules, in Hegner's leaf beetle
studies, 50–51
potency property, stem cells, 76–78
preformation (maternal inheritance), 68–
69, 72, 92
primordial germ cell–like cells, 100–101,
102
primordial germ cells: in annelids, 90; in
crustaceans, 92; differentiation processes, 68–70, 72, 105; as germline
initiation, 2–3; Hargitt's experiments,
45–46, 63; Hegner's theorizing, 50–51,
54; in humans, 101–2; maternal inheritance process, 68–69; in mice, 101; in
sea urchins, 94; in tunicates, 88–89.
See also germ cells *and related entries*

Pristina leidyi, natural transdifferentiation,
90–91
progeny standard, 98–99, 100, 101

Rana pipiens studies, Bournoure's, 55–56
rat experiments, 46, 110
rDNA technology, 110
recombinant DNA (rDNA) technology, 110
reconstituted ovaries, mice experiments,
101
regeneration. *See* germline regeneration,
overviews; germline regeneration,
traditional perspectives
Roux, Wilhelm, 27, 95
Rubaschkin, W., 48

Saitou, Mitinori, 99–101, 102
salamanders, 5, 45–46
Sasaki, Kotaro, 102
sea urchins, natural transdifferentiation,
93–94
Seaver, Elaine, 89–90, 98–99
selective staining criteria problem, germ
cells, 48–49
self-renewal property, stem cells, 76
sex cells. *See* germ cells *and related entries*
Simkins, Cleveland, 48, 58
Simpson, George Gaylord, 37–39, 57
somatic cell nuclear transfer, 95–96
somatic cells: and epigenesis mode, 70, 71;
gene therapy debates, 115–22; genome
editing debates, 114–15; in *Hydractinia* experiments, 83; identification
assumption, 3–4, 62–65, 103–4; in
models of regeneration, 74, 75, 76; in
Morgan's planarian experiments, 84;
stem cell question, 78
somatic cells, historic theorizing: Hargitt's
embryological experiments, 43–47;
in Simpson's biology textbook, 37–39;
Weismann's germ plasm concept, 22–
23, 25–26, 52–53; Wilson's cell studies,
28, 29–30
somatic cells, transdifferentiation
evidence: in annelids, 90–91; in
crustaceans, 92; in mice, 100–101;
overview, 103; in tunicates, 88–89. *See
also* induced pluripotent stem cells;

Printed and bound by CPI Group (UK) Ltd, Croydon, CR0 4YY

27/10/2024

14580400-0001

induced transdifferentiation; Weismann Barrier
somatic cells from mice, human transdifferentiation studies, 102–3
soma-to-germ transitions, 121
specification. *See* differentiation processes, overview
spermatozoa, defined, 2
spermatozoa, documentation history, 11–13
Splicing Life report, 114
sponge cells, in pluripotency regeneration model, 79–82
starfish, 93
stem cells: in annelids, 91; inn pluripotency model of regeneration, 79–85; properties summarized, 76–78, 96; somatic cell question, 78. *See also* induced pluripotent stem cells
stroma, as germ cell location, 45–46
syncytia, 69

Takahashi, Kazutoshi, 96–97
Takamura, Katsumi, 88
TALENs (transcription activator-like effector nucleases), 110–11, 113
ß-thalassemia, 115–16
totipotency ability, stem cells, 78
traditional model, germline regeneration, 74
transcription activator-like effector nucleases (TALENs) technology, 110–11, 113
transcription factors, 96–97, 101–2
transdifferentiation model, germline regeneration, 74, 74, 85. *See also* induced transdifferentiation; natural transdifferentiation
transmission genetics, defined, 36
Tschermak-Seysenegg, Erich von, 35
Tudor mutations, 133n2
tunicates, natural transdifferentiation, 85–89

unipotency ability, stem cells, 76

van Beneden, Edouard, 13–14
virus delivery, gene therapies, 112–13

Waldeyer, Heinrich Wilhelm Gottfried, 13–14, 129n5
Weintraub, Harold, 96
Weismann, August (his theorizing): criticisms of, 26–27, 28; germ plasm and generation connection, 70; germ plasm architecture, 19–22, 52–53; Hegner's conflation of, 54–55; hydrozoan studies, 44–45, 82–83; insect observations, 51–52; mouse tail experiments, 18; nuclear division processes, 21–22, 25–26; role of organismal development, 18–19, 20–21, 23–25, 28; Wilson's responses to, 27–31
Weismann Barrier: as cornerstone of early genetics theorizing, 33–41; genetics vs. embryology perspectives, 34–35, 43; in genome editing debates, 117–22; historic assumption, 6–7, 10, 41, 62, 73, 105; and models of regeneration, 73–75, 78; as normative claim, 42, 105; permeability evidence, 105–6, 119; Simpson's incorporation, 37–39; in Wilson's theorizing, 30–31, 34, 37. *See also* induced transdifferentiation; natural transdifferentiation
Whitman, Charles Otis, 32
Wilson, Edmund Beecher, 27–31, 37

xenogenic reconstituted ovaries (xrOvaries), 102–3
xrOvaries (xenogenic reconstituted ovaries), 102–3
X-SCID disease, 112–13

Yamanaka, Shinya, 96–97, 99, 102
Yamashiro, Chika, 102
Yamashita, Yukiko, 70–71
Yoshida, Keita, 88

ZFNs technology, 110–11, 113
zinc finger nucleases (ZFNs) technology, 110–11, 113
zygotes, division processes, 69–70